DISCRIMINATION OF GENERA OF EUPLECTINI

OF NORTH AND CENTRAL AMERICA

(Coleoptera: Pselaphidae)

Discrimination of Genera of Euplectini of North and Central America

(Coleoptera: Pselaphidae)

by Albert A. Grigarick and Robert O. Schuster

UNIVERSITY OF CALIFORNIA PRESS
Berkeley • Los Angeles • London

UNIVERSITY OF CALIFORNIA PUBLICATIONS IN ENTOMOLOGY

Advisory Editors: H. V. Daly, J. A. Powell, J. N. Belkin, R. M. Bohart,
D. P. Furman, J. D. Pinto, E. I. Schlinger, R. W. Thorp

Volume 87

Issue Date: March 1980

UNIVERSITY OF CALIFORNIA PRESS
BERKELEY AND LOS ANGELES, CALIFORNIA

UNIVERSITY OF CALIFORNIA PRESS, LTD.
LONDON, ENGLAND

Library of Congress Cataloging in Publication Data

Grigarick, Albert A
    Discrimination of genera of Euplectini of North
and Central America (Coleoptera, Pselaphidae)

    (University of California publications in entomology;
v. 87)
    Includes bibliographical references and index.
    1. Pselaphidae—Classification.   2. Insects—
Classification.   3. Insects—North America—Classifica-
tion.   4. Insects—Central America—Classification.
I. Schuster, Robert O., joint author.   II. Title.
III. Series: California. University. University of
California publications in entomology ; v. 87.
QL596.P8G74   595.7'64   79-63591
ISBN 0-520-09609-6

ISBN 0-520-09609-6
LIBRARY OF CONGRESS CATALOG CARD NUMBER: 79-63591

Contents

Acknowledgments

The assistance and cooperation of a great many persons who searched for specimens, answered questions, and provided examples for study made this revision possible. We gratefully acknowledge the use of specimens from the following collections, and sincerely thank their staffs: Academy of Natural Sciences of Philadelphia (ANSP), Drs. M. G. Emsley, D. C. Rentz, and W. W. Moss; American Museum of Natural History (AMNH) Dr. L. W. Herman, Jr.; British Museum (Natural History) (BMNH), Mr. M. E. Bacchus; California Academy of Sciences (CAS), Drs. P. H. Arnaud, Jr., and D. H. Kavanaugh; Carnegie Museum (CM), Dr. G. E. Wallace; Cornell University (CU), Dr. L. L. Pechuman; Field Museum of Natural History (FMNH), Drs. J. B. Kethley, R. Wenzel, and M. Prokop; Museum of Comparative Zoology (MCZ), Dr. P. J. Darlington, Jr., and Mrs. J. C. Scott; Museum National D'Histoire Naturelle, Paris (MNHN), Dr. J. J. Menier and Mme. A. Bons; National Museum of Natural History (USNM), Dr. P. J. Spangler.

Many interesting specimens were made available from the private collections of Dr. D. S. Chandler, Mr. H. R. Steeves, Jr., and Mr. K. Stephan. Dr. J. A. Wagner assisted in locating specimens in the Park Collection at the Field Museum of Natural History.

Ms. Karen Calden Jones illustrated the dorsal aspect of some genera, and these figures are so initialed. The remaining illustrations were drawn by the junior author.

INTRODUCTION

The genera of very small beetles formerly encompassed by the tribes Euplectini and Trichonychini are discussed in this monograph. The separation of these tribes was based on the degree of reduction of the inner (accessory/secondary) tarsal claw (Raffray, 1890), a character (Plate 78; figs. 1,2) which is difficult to detect or to interpret and which, for the most part, fails to indicate fundamental relationships. The formal combination of these tribes to Euplectini *sensu latiore* was made by Park (1942), but earlier workers such as Brendel and Wickham (1890) and Casey (1893) were of the same opinion. Our examination of nearly 80 genera neither supports a division based on the accessory tarsal claw nor provides alternative characters in support of a Euplectini-Trichonychini grouping of genera.

The tribe Euplectini now consists of more than 120 genera and is distinguished from other tribes of pselaphids by the following characteristics: (1) the mesotrochanter (75;2) is short and obliquely articulated with the femur, so that the femur is relatively close to the coxa; (2) the metathoracic coxa (75;1) is transverse except at the inner face, where it becomes conically produced for articulation with the trochanter; (3) the tarsus is predominantly three-segmented, with the first segment being very small (this tarsomere is absent in two genera); (4) the maxillary palpus (75;3) is four-segmented, the first being very small, the second narrow basally and enlarged apically, the third smaller than the second and of varied shape, and the fourth segment largest, more or less oval, with a small palpal cone and two to four parallel tubular setae appressed on the dorsal surface; (5) the abdomen generally is elongate and has five visible tergites, of which the first four are distinctly margined.

The Euplectini are widely distributed throughout the world, with the Americas being rich both in genera and species. This study was undertaken to find definitive morphological characters for generic discrimination and to provide additional information for assessing the presumed phylogenetic relationships of the genera. Because the desired information was not available in the literature and could not be seen on point-mounted specimens, slide preparations of whole or dissected specimens were prepared for each genus. Whenever possible, the holotype of the type-species of each genus was examined to insure valid association of observations and genera. Information obtained from examination of the slide preparations indicated the need to synonymize a few genera and to propose yet others. It also showed the existing subtribal classification to have little or no phylogenetic significance, or at least to be inapplicable to the phylogeny we have proposed.

1

Our study was confined to North and Central America and associated islands, primarily because of the availability of type specimens. Although limited in scope, a reasonable overview of the Euplectini has emerged from this study of 80 genera, or about two-thirds of the worldwide genera included in the broad definition of the tribe.

A minimum of 30 morphological characters with two to five states per character is provided for each genus. This information is presented by means of tabular keys, a bracket key, text, and selected illustrations. Numbers referring to illustrations will always be in pairs, the plate number followed by a semicolon and the figure number. The characters used in the keys are equally applicable to both sexes. Sex-limited characters are often illustrated or mentioned in text and, while not comparative in the sense of characters used in the keys, still have significant value in the recognition of taxa.

Genera can be most rapidly identified by utilizing the bracket key. This identification should then be confirmed by comparison of characters in the tabular keys (Tables 3, 4 and 5) and illustrations for that genus. Phylogenetic relationships are proposed, diagrammed, and discussed on the basis of the aggregate of characters listed in Table 1.

MORPHOLOGIC AND PHYLOGENETIC CONSIDERATIONS

The tribe Euplectini has been divided into numerous subtribes by nearly as many authors as there are divisions. Various characters such as the antennal configuration, surface carinae, reduction of the accessory tarsal claw, proximity of mesocoxal cavities, and the separation of metacoxae have been used to separate subtribes. While many of these characters have been useful in distinguishing genera, relatively few are of value in showing relationships among groups of genera having many other characteristics in common. For example, the elongate frontal rostrum of the subtribe Rhinoscepsina held together two completely unrelated genera, *Rhinoscepsis* and *Morius*. We have not used subtribe designations, but have divided the genera into three groups based on common characters of the mesosternum that we believe to be fundamental to structure and habit. Within these three large groups there are some aggregations of genera of close affinities, but there are other genera that stand by themselves or in pairs that are difficult to associate other than by the mesosternal characters.

The eventual consideration of the remainder of the tribe will indicate the extent to which the present three groups can cope with the added diversity. Subtribal designations could be assigned in the future to the groups or some modified grouping, but it would in no way resemble the existing subtribe classification. For instance, ten genera in Group A (those without foveae associated with the antennal insertions) were formerly placed in the six subtribes Euplectina, Trimiina, Trisignia, Bibloporina, Rhinoscepsina, and Panaphantina. The remainder of Group A includes most of the subtribe Trogastrina and genera from Bibloporina, Euplectina, Rhexina, and Rhexinia.

The presence and position of specific foveae or their absence have played a very important role in our discrimination of genera and proposed phylogeny. The potential value of a foveal system in a scheme of classification of Pselaphidae was proposed by Park (1942:24-28). He stated that the topographic position of many foveae in Pselaphidae tends to indicate that they have arisen where sutures coalesce. Therefore, he believed the presence of these integmental invaginations to be an indirect argument for the structural consolidation of the body. This consolidation is an important characteristic of pselaphids that distinguishes them from the less consolidated, and generally presumed to be more primitive, staphylinids. Park hypothesized the following sequence of evolution: (1) developed sclerites bounded by clean sutures; (2) sclerites becoming consolidated, the sutures vestigial and their more resistant intersections or areas persisting as foveae; and (3) eventual disappearance of foveae, leaving a solid sclerotized piece. He

TABLE 1. Characters selected as ancestral or derived, and their assigned values for phylogenetic placement of genera of Euplectini

ANCESTRAL		DERIVED	
Head	value	*Head*	value
Antennal foveae present	.5	Antennal foveae absent	0
Ventral median carina present	.5	Ventral median carina absent	0
Thorax		*Thorax*	
Median pronotal impression present	.5	Median pronotal impression absent	0
Basal pronotal impressions present	.5	Basal pronotal impressions absent	0
Elytron with subhumeral foveae	.5	Elytron without subhumeral fovea	0
Elytron with antebasal fovea	.5	Elytron without antebasal fovea	0
Prosternum with median carina	.5	Prosternum without median carina	0
Procoxal foveae paired	1.0	Procoxal foveae absent	0
Anterior prosternal fovea paired	.5	Anterior prosternal foveae absent	0
Mesosternum with median carina	.5	Mesosternum without median carina	0
Median mesosternal foveae present	—	Median mesosternal foveae absent	0
Paired	3.0		—
Single-forked	2.7		—
Single	1.0		—
Lateral mesosternal foveae paired	3.0	Lateral mesosternal foveae paired-forked	2.0
Lateral mesocoxal foveae paired	2.0	Lateral mesocoxal foveae absent	0
Promesocoxal foveae paired	.5	Promesocoxal foveae absent	0
Metasternal foveae present	—	Metasternal foveae absent	0
Paired	1.0		—
Single	.5		—
Abdomen		*Abdomen*	
Tergite I with setate basal depression	.5	Tergite I without setate basal depression	0
Sternite II with foveae	—	Sternite II without foveae	0
Four foveae	1.0		—
Two foveae	.5		—

selected a number of genera in the family to examine the foveae of the meso- and metasternal areas. The genus *Sonoma* in the tribe Faronini has the greatest number of foveae, and this tribe is generally listed first in catalogues and considered to be the most primitive by pselaphidologists such as Casey, Bowman, Jeannel, and Park. There is a paucity of information on the phylogeny of this very large family of very small beetles.

In the course of our study of the genera within the tribe Euplectini, several evolutionary trends appeared to support Parks' hypothesis. The characters selected to show these trends and to serve as a basis for our proposed phylogeny for Euplectini are presented in Table 1. These characters were selected because of their inter- and intraspecific constancy and their distinctness. Most of the characters are either present or absent and require little interpretation. The state of these characters for each genus is presented through Tables 2 to 5. If a character is uniformly present with the genera of a group, it is so stated in the introductory material of Groups A, B, and C rather than being placed in the tables.

Tables 3 to 5 include some characters that are not defined as ancestral or derived because they may not indicate evolutionary trends, may show more variability within a genus, or are more difficult to detect and interpret. These supplementary characters, in combination with those of Table 1, have been useful in defining genera and clusters of genera. Only 19% of the genera of this tribe have five or more species; 68% are based on one or two species. The full complement of species of some genera was not always available for dissection and examination. The constancy of characters previously mentioned was determined in the course of study of the total species of some of the larger genera in Groups A, B, and C, including *Oropus* (26 species), *Oropodes* (5 species), *Actiastes* (9 species), and *Actium* (35 species).

The numerical values assigned to the characters listed in Table 1 were made on the basis of our estimation of their overall importance in showing evolutionary trends and their constancy within species of the tribe. The evolutionary trend, change from complex body areas to less complex areas by way of reduction in numbers of foveae, fewer impressions, and less carinae, is discussed above in a general way with respect to Parks' ideas. These trends are also taken up later character by character and discussed on the basis of their frequency within the three main groups and the ways change may have come about. The venter of the beetle, and particularly the mesosternum, was given the highest priority in selecting the numbers of characters and the numerical values assigned to these characters. The prosternum and metasternum were next highest in the rating and were considered of about equal value. Sternite II was also believed to be quite important. It was evident that the rate of evolutionary change varied with the different genera, so that a genus with a majority of ancestral characters could also have derived characters. In general, the progression from an ancestral to a derived condition followed the pattern of reduction, consolidation, and elimination. Since the absence of a primary character (e.g., median mesosternal foveae) could come about in several ways through parallelism, secondary characters were sometimes relied upon to indicate the probable evolutionary path and assist in the assignment of a genus to a particular group.

The genera of Euplectini we examined were assigned to three groups, based on many characteristics in common but separated by the characters listed in Table 2. The relationship and relative sizes of these groups (number of genera) are presented in Diagram 1. The generic assignments to Groups A, B, and C were based on the summation of values given to the foveae of the mesosternum (Table 1). An introductory section with each group discusses the proposed phylogeny of that group and gives the characters in common.

Diagrams 2 to 6 depict the relationships of genera assigned to the groups. The genera are positioned on these diagrams vertically, on the basis of a summation of the values of ancestral or derived characters listed in Table 1. The uppermost genera are considered to be the most advanced, but there is no intent for this vertical placement to be a direct temporal assignment. The horizontal placement indicates relatedness within the particular group, as shown by the numbers and letters associated with the horizontal separation. The characters represented by the numbers and letters are given in the text to that diagram. The most closely related genera (clusters) are shown by an absence of

TABLE 2. Tabular key to groups of Euplectini based on the foveae of the mesosternum[1]

FOVEAE	PAIRED	FORKED	SINGLE	FORKED	ABSENT
(Group A)					
Lateral mesosternal	X	—	—	—	—
Median mesosternal	X[a]	—	(X)	(X)	(X)[b]
Promesocoxal	(X)	—	—	—	X
Lateral Mesoscoxal	X	—	—	—	—
(Group B)					
Lateral mesosternal	X	X	—	—	—
Median mesosternal	X[a]	—	(X)	(X)	—
Promesocoxal	(X)	—	—	—	X
Lateral mesocoxal	X	—	—	—	—
(Group C)					
Lateral mesosternal	X	—	—	—	—
Median mesosternal	—	—	X	—	(X)
Promesocoxal	—	—	—	—	X
Lateral mesocoxal	X	—	—	—	(X)

1. X—foveae present
   (X) —less frequent occurrence of multiple condition
   a—bases or sides of foveae may be joined
   b—if absent, antennal foveae are present

numbers or letters in their horizontal placement. The table associated with each group (tabular key to that group) and the common characters listed for the group, along with the values assigned in Table 1, can be used to check our vertical and horizontal assignment of genera in this paper or to compare these genera with others. Tables 3, 4, and 5 are not designed to show phylogeny. The relationships within clusters of genera are considered under the discussions of those genera.

MORPHOLOGICAL FEATURES USED TO DISCRIMINATE GENERA

Whole, point-mounted specimens should be carefully examined with respect to depressions, carinae, and foveal openings, but it will nearly always be necessary to partially clear and slide-mount specimens to determine the position of many foveae as well as other characteristics essential for generic discrimination.

Head

The shape of the head varies considerably in dorsal aspect, ranging from nearly oval, triangular, or trapezoidal to broadly expanded anterolaterally, as in *Malleoceps* (67;2), and extended into a rostrum in *Morius* (26;3) and *Rhinoscepsis* (2;2). Foveation of the head provides many distinguishing features. The vertexal fovea are present in all genera except *Trimiosella*, but a few are reduced (75;4) and nearly closed. The vertexal fovea connect to the tentorium, which terminates ventrally as the gular foveae. The gular foveae are always present and have either two separate or a common median opening

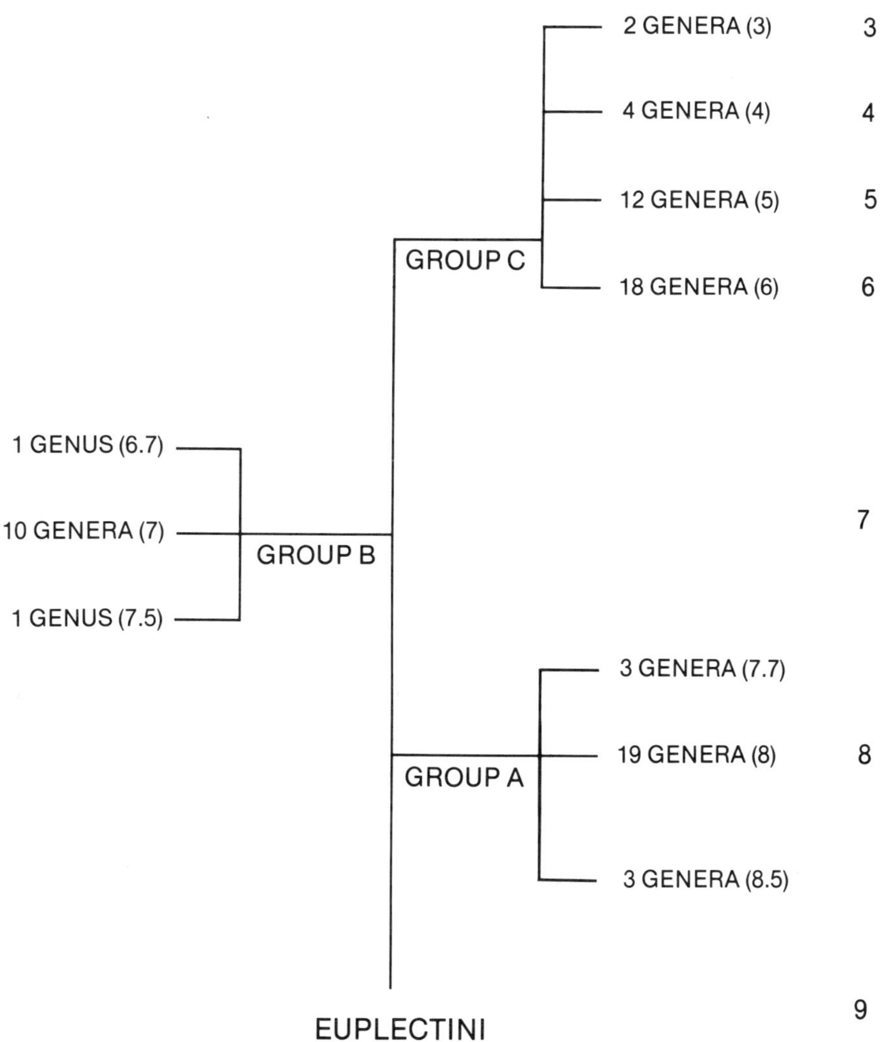

DIAGRAM 1. Placement of genera in Groups A, B, and C on the basis of a summation of values assigned to ancestral or derived states of foveae of the meso-sternum (Table 1)

ventrally. Other foveae are associated with the antennal insertions of 15 genera. These paired foveae open on the head dorsally (75;4) and/or laterally. Both the dorsal and the lateral pair may be present (*Rhexidius*—14;1), or either pair may occur separately (*Eurhexius*—19;1, *Tuberoplectus*—12;1). Sometimes the foveae extend to and become a part of the tentorium (*Rhexinia*—21;1), or they terminate blindly (*Oropus*—13;1, *Hatchia*—23;1). Sometimes the walls of these lateral foveate areas are not of the typical spiral construction but are rather large and could be considered as sinuses.

Paired foveate areas are present near the base of the antennae in 60% of the genera of Group A and are limited to that group. Six of the genera with extra foveation of the head also have antennal segment I elongated from two to four times the length of segment II. The only other genus showing this degree of segment elongation is *Rhinoscepsis* (2;2). A ventral median carina occurs longitudinally on the head of some genera of Groups A, B, and C in 40%, 17%, and 9% respectively.

Pronotum

A median longitudinal impression or sulcus may occur on the pronotal disc. The length and depth of this impression varies considerably among different genera. It occurs in 76% of Group A, 58% of B, and 14% of C. Basal depressions occur on the pronotum of some genera in Groups A and B only. These depressions are found on only 12% of the genera of A, but are on 67% of the genera of B. They occur with or without the discal impression, but predominantly with it. A transverse, biarcuate sulcus nearly always extends between the lateral pronotal foveae. These foveae are usually large and distinctly dorsal, but are lateral and extremely reduced in size for a few genera of Group C.

Elytra

The length and particularly the width of the elytra in relation to the prothorax and abdomen vary to alter the general appearance of species, but no consistent relationship of shape that would characterize related groups of genera was observed. The sutural fovea (79;3,4) and stria are consistently found on the elytra of all genera except *Hanfordia*. The number of discal or antebasal foveae range from three to zero, with the majority having one. The absence of discal foveae is generally associated with a reduction of foveae in other areas. The subhumeral fovea is present in 80% of Group A, 75% of B, and 49% of C. The epipleural stria is usually present even though the subhumeral fovea may be absent.

Prosternum

When present, the procoxal foveae are always paired, well separated or close, and vary in size from large to minute. Park (1942) called these foveae "lateral prosternal foveae," but remarked that they were not always lateral and were always associated with the procoxae, so the term procoxal foveae seems more appropriate. On occasion only the outer foveal ring remains on the closed surface as evidence of their former position. The procoxal foveae are present in all genera of Group A except the highly advanced genus *Aboeurhexius*, in all genera of B, and in 60% of C. The procoxal foveae are considered the most useful character on this segment for determining phylogenetic relationships.

A pair of anterior prosternal foveae occurs in only five genera, four in Group A and one in Group C. Some of these genera are considered advanced, and there does not seem to be a consistent pattern to the occasional presence of anterior prosternal foveae. The longitudinal median carina occurs on three to four genera of each group. There is no correlation of this carina with the presence or absence of procoxal or anterior prosternal foveae.

Mesosternum

Foveae in the preplectoid area were not found in the genera of Euplectini examined, although distinct depressions in this area were observed for a few genera.

A pair of lateral mesosternal foveae (76;1,2) are present in all genera. These foveae may vary in width, length, and angle of intrusion, but they are always prominent and are the only foveae of the mesosternal area that have not shown an evolutionary trend toward reduction in this tribe. This presumably indicates that they have an important function which, because of their position, is probably associated with locomotion. Group B is established for those genera in which the lateral mesosternal foveae are forked. The junction of the fork in relation to the surface opening varies to the degree that a few genera appear to have a pair of foveae on each side (*Thesiectus*—36;2), but the junction of the inner surface of the fork was clearly visible for most genera of Group B. It would be interesting to determine if some difference in locomotive pattern of this group is associated with the forked condition of the lateral mesosternal foveae.

The median mesosternal foveae may be paired (76;1 and 77;1), forked with a single opening, single (77;2,3,4), or absent. The most ancestral or paired condition is prevalent in Groups A and B. The single-opening, forked condition occurs in three genera of Group A—*Rhexiola* (22;1), *Hatchia* (23;4), and *Trigonoplectus* (7;4)—and the genus *Trichonyx* (30;4) of Group B. The rather skewed median mesosternal foveae of *Trisignis* (4;5) of Group A appear as a wide single median fovea, but close examination shows a junction of two foveae along the median side. In the genus *Aboeurhexius* (25;5) of Group A these foveae are absent. The median mesosternal fovea of genera of Group C is usually single, but is absent in about one-quarter of the genera. The occasional forked condition and lateral junction in genera of Groups A and B would indicate that the single fovea is a condition derived from the coalescing of paired foveae. The loss of mesosternal foveae may have followed the coalescence to a single fovea or could have developed directly from the paired condition. The former is believed to have occurred more frequently.

A pair of promesocoxal foveae are found only in the genera *Thesium* (1;2), *Tubero-plectus* (12;7), and *Barroeuplectoides* (11;6) of Group A (considered among the more ancestral genera), and *Euplecturga* (32;7) of Group B.

A pair of lateral mesocoxal foveae (76;1,2) is consistently found in all genera of Groups A and B and 80% of C. It is frequently the largest pair of foveae, and is absent (or extremely reduced) only in highly advanced melboid (71;5) genera. Park (1942) reported the lateral mesosternal foveae to connect to the arm of the mesosternal furca of the endoskeleton.

The mesosternal surface area between the lateral mesosternal foveae and mesocoxal cavities exhibits a pattern that is very characteristic (77;1-4). The shape, type of reticulation, and setal pattern displayed on this area are frequently of value in showing generic relationships. The mesocoxal cavities are completely separated by the extension of median processes or are open to varying degrees. The more advanced genera of Groups A and B predominantly show the open condition, but this is not consistent. All genera of Group C have open mesocoxal cavities.

Metasternum

Three, two (77;3), one (77;2), or no (77;1) foveae occur posterior to the mesocoxa. These foveae vary in size and spacing, nearly always appear to open on the suture separating the mesosternum from the metasternum, and are usually directed medioanteriorly. These foveae were found in 76% of the genera examined. Park (1942) classified these foveae as posterior mesocoxal foveae or median metasternal foveae, but did not clearly separate the two types. He apparently considered all single median foveae and paired foveae close to the median to be metasternal foveae. Because the degree of separation is not distinct, we are arbitrarily considering them all to be metasternal foveae. When present, the paired condition predominates (76;1,2 and 77;3,4). *Aboeurhexius* (25;5) in Group A has a single median fovea, and *Rhinoscepsis* (2;1), in the same group, has three—a pair of lateral foveae and a single median fovea. The single condition does not occur in Group B and is found only in *Allobrox* (53;2) of Group C. The evolutionary trend of paired metasternal foveae appears to have taken two courses. They either became reduced in size until they were absent, or were directed medianly until they fused to a single median fovea. No minute remnant of a single median fovea has been observed, but guard setae (frequently associated with foveal openings) have been found in the mid-position without a fovea, and it is assumed that the absence of metasternal foveae could also result from the loss of the single median fovea. Metasternal foveae are present in 60% of the genera of Group A, 74% of B, and 86% of C.

Legs

Many of the modifications of the various segments of the legs are a product of sexual dimorphism and are not used to distinguish genera. However, an area of the profemur of both sexes of many genera is carinate and/or has specialized setae (78;3,4) assumed to be of sensory function. In some genera of Groups A and C these modifications of the profemur are in the form of distinctive large circular pits (*Thesiastes*—48;6, *Trimioplectus*—8;1), and the depressions in the profemur of *Melba* (78;4) contain ramified setae. The presence of the various sensory modifications occurs in 68% of the genera of Group A, 17% of B, and 71% of C. No consistent evolutionary trend was established, other than that the sensory areas are usually similar for related genera. The metacoxae are usually contiguous; however, the metasternal extension between five of the genera of Group C are quite broad (*Biblomelba*—58;5). Three tarsal segments are typical for all genera except *Tuberoplectus* and *Barroeuplectoides* (11;7), which have only two.

Abdominal Tergites

The first two tergites, and usually the third, are hidden below the elytra of point-mounted specimens. Segments four to eight are visible and are designated I to V. The relative lengths of these segments are consistent for both sexes at the generic level, and this characteristic is recorded for segments I to IV. Tergite I is the longest in about 60% of the genera; these four tergites are subequal in approximately 30%; and tergite IV is the longest in most of the remainder. No consistent pattern of tergite lengths is associated with groups.

Tergites I to IV frequently have a median basal depression and/or longitudinal carinae which are usually distinctive at the generic level for the species examined. The degree of indentation varies from barely discernable to obvious. Specialized setae are nearly always present in the more obvious depressions, and this setate condition was used in generic characterization. The basal depressions sometimes have foveae at their lateral margins (13;8). Longitudinal basal carinae vary in length and shape from narrow and long (*Tomoplectus*—42;8) to short and blunt (*Latomelba*—64;8). When the basal depression is absent, the carinae are quite obvious. The lateral margins of the basal depressions sometimes appear as carinae (*Thesiastes*—48;5), but are not considered as such unless the line is raised and extends beyond the depression (*Trigonoplectus*—7;6). The number of segments with basal carinae and/or basal setate depressions varies, but when present these carinae and/or depressions are always on tergite I, and when on more than one segment, always occur on the previous segment. There are basal setate depressions on 68% of the genera of Group A and carinae on 48%. In Group B, 33% have basal setate depressions and no genus has basal carinae. In Group C, only 17% of the genera have basal setate depressions, but 80% have basal carinae.

Tergites I to IV are laterally margined, and the anterior part of the margin of tergite I is usually reticulate. A projection of the elytron fits this roughened area, and Park (1963) has suggested an auditory function for the structure. Setate foveae occur at the antero-lateral margins of tergites I to IV for *Oropus* (13;8), *Hexirhexius*, *Rhexidius* (14;6), and *Euboarhexius*.

Abdominal Sternites

The first sternite (I) usually visible is the third true segment and may be somewhat obscured on point-mounts. Sternite II is prominent and may possess depressed areas associated with one (*Aboeurhexius*—25;9) or two pairs (*Trichonyx*—30;8, *Pycnoplectus* —9;6) of foveae, or may be without modifications. The number of fovea (two or four) or their absence is used in defining genera. In all three groups, 50% to 60% of the genera contain two foveae on sternite II. Four foveae occur in 25% of the genera of Group A, 41% of B, and only 6% of C. The absence of foveae is apparently the derived condition, since they are absent in Groups A and B only 8% of the time, but are missing 46% of the time in Group C.

Sex-limited Structures

Males frequently have sex-limited structures which occur, with few exceptions, on the abdominal segments and are distinctive of each species of each genus. These structures typically consist of a lamellate integumental projection and a number of specialized setae, but for a few genera (e.g., *Oropus*, *Rhexidius*, *Hexirhexius*) are shallow porous depressions filled with microsetae. The fine-structure of these male characters was not investigated. However, the porous integument and the area protected by integumental lamellae or specialized setae suggest that these structures may be concerned with the production of pheromones.

Additional Characters

Additional characters that should be of phylogenetic significance, but were not presently comparable, include: (1) the types and distribution of sensory setae on the antennae and ventral surface of the head; (2) ultrastructural morphology of the setae associated with foveae, and the specialized setae of the profemur; (3) the relative widths of the femora; and (4) the number of tergites of the male, and orientation of the genitalia. These characters are not yet known for various reasons: because the types of some genera lacked parts, or the characters could not be discerned from the limited examples available; or the sexes were unknown; or institutional policy prevented suitable specimen preparation. Some reorganization of the proposed classification is anticipated when these and other criteria become available in comparative form.

GROUP A

Of the 25 genera placed in this group, 18 are found north of Mexico and 12 in Central America or the islands north of South America. *Thesium*, *Rhinoscepsis*, *Rhexius*, and *Euplectus* are reported from North, Central, and South America, with *Euplectus* being cosmopolitan in distribution. Many new genera have been established from species of *Euplectus* in the past, and its distribution may prove to be more limited in the future, upon reexamination of its remaining species.

Genera are assigned to Group A based on the foveae of the mesosternum (Table 2), primarily the nonforked lateral mesosternal foveae and the paired median mesosternal foveae or median mesosternal fovea(e) showing evidence of having been paired (see discussion in previous section). The most primitive genera (by the selected characteristics of Table 1) are found in this group. *Thesium* has many of these ancestral characters, and we consider it to be the most generalized genus in the tribe.

A phylogenetic arrangement of genera in Group A is proposed in Diagrams 2 and 3. Relationships are discussed as each genus is considered individually.

A tabular key to genera placed in Group A, based on 29 characters for each genus, is given in Table 3. (The genera designated with an asterisk on the right margin may not always be separated by the 29 characters listed.) The text, figures, and bracket key should be consulted for additional characters. In addition to the common mesosternal characters, all genera of Group A have contiguous metacoxae and elytra with antebasal foveae.

Thesium
Casey, 1884:94, 117
(Plate 1, figs. 1-7; Plate 79, fig. 1)

Type-species: *Euplectus cavifrons* LeConte, 1863. Smithsonian Misc. Collections VI, no. 167:28. New Orleans Designated by Bowman, 1934:144.

Distribution: 15 species have been described in this genus from South, Central, and North America.

Diagnosis: Members of this genus are relatively robust but do not exhibit particularly distinctive external features. *Thesium* is readily separated from other genera in the group by possessing a median prosternal carina (1;1); two basal depressions on the pronotum (1;3); and no foveae near the base of the antennae. The mesosternal area (1;2) is unique for all genera considered because of the uniform size and placement of all the main foveae; a median longitudinal carina on the surface area; and a broad extension separating the mesocoxae. This mesosternum is considered to be the most generalized, and the genus is placed lowest in the phylogenetic arrangement of Diagram 2. Specimens studied were from the Park collection.

13

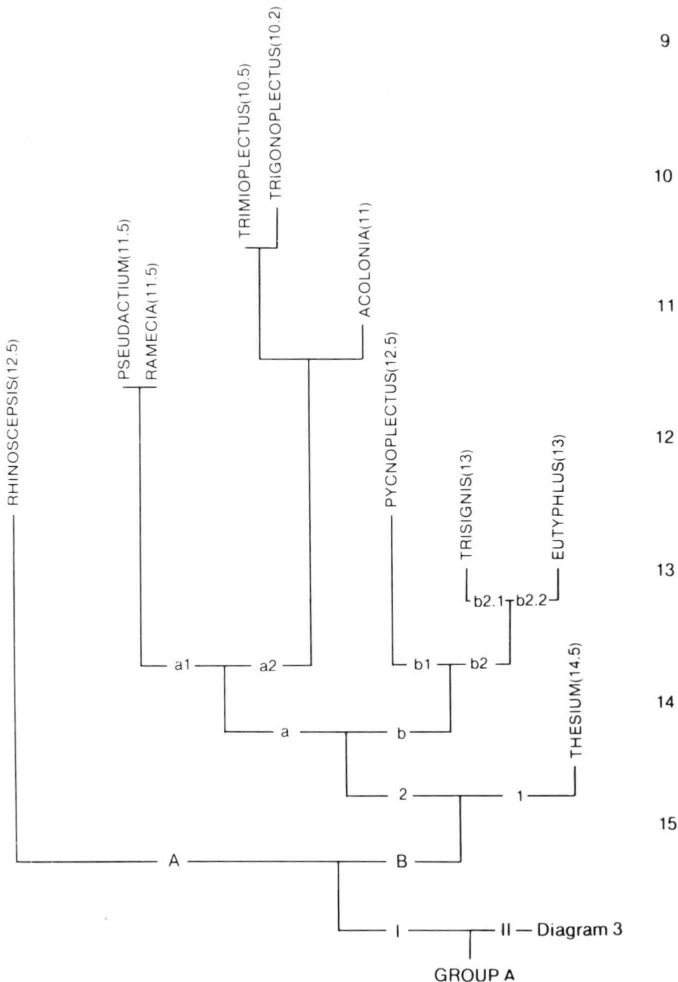

I. Head without antennal foveae
  A. Metasternum with 3 foveae
  B. Metasternum with 2 or no foveae
    1. Mesosternum with median carina; with promesocoxal foveae
    2. Mesosternum without median carina, without promesocoxal foveae
      a. Pronotum without median depression
        a1. Metasternum with 2 foveae
        a2. Metasternum without foveae
          a2.1. Elytron without subhumeral fovea
          a2.2 Elytron with subhumeral fovea
      b. Pronotum with median depression
        b1. Pronotum with basal depressions
        b2. Pronotum without basal depressions
          b2.1. Head with ventral median carina
          b2.2. Head without ventral median carina

DIAGRAM 2. Assignment of genera to Group A, part I, on the basis of a summation of values of ancestral or derived characters listed in Tables 1, 2, and 3

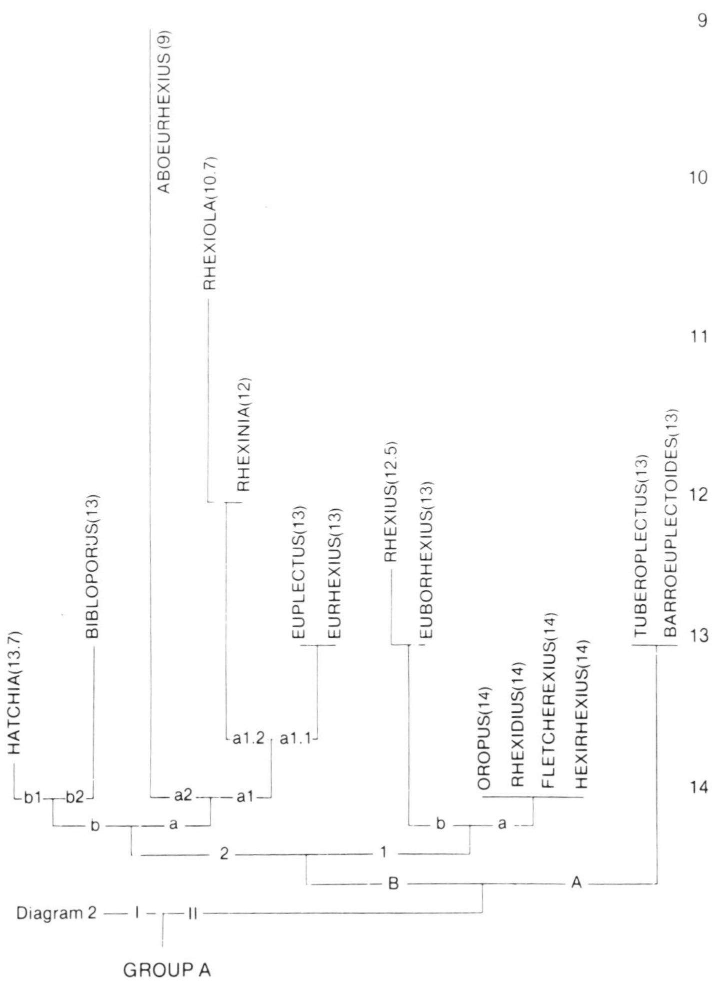

II. Head with antennal foveae
   A. Mesosternum with promesocoxal foveae
   B. Mesosternum without promesocoxal foveae
     1. Sternite II with 4 foveae
       a. Metasternum with 2 foveae
       b. Metasternum without foveae
     2. Sternite II with 2 or no foveae
       a. Prosternum without median carina
         a1. Prosternum with procoxal foveae
           a1.1. Head with ventral median carina
           a1.2. Head without ventral median carina
         a2. Prosternum without procoxal foveae
       b. Prosternum with median carina
         b1. Head with ventral median carina
         b2. Head without ventral median carina

DIAGRAM 3. Assignment of genera to Group A, part II, on the basis of a summation of values of ancestral or derived characters listed in Tables 1, 2, and 3

 University of California Publications in Entomology

TABLE 3. Tabular key to the genera of Group A

GENUS	HEAD					THORAX															TERGITES								S.
	Frontal rostrum	Antennal foveae opening dorsally	Antennal foveae coalesced dorsally	Antennal foveae opening laterally	Ventral median carina	Median pronotal impression	Lateral longitudinal impressions	Basal pronotal impressions	Elytron with subhumeral fovea	Median carina on prosternum	Anterior prosternal foveae	Procoxal foveae	Median carina on mesosternum	Median mesosternal foveae paired	Median mesosternal f. single, forked	Promesocoxal foveae	Metasternal foveae paired	Metasternal fovea single	Profemur with sensory modification	Tarsi with 2 segments	Setate basal depression on I only	Setate basal depression on I–II	Setate basal depression on I–IV	Basal carina on I only	Basal carinae on I–II	Length subequal	I longest	IV longest	Foveae on sternite II
Thesium	−	−	−	−	−	+	−	+	+	+	−	+	+	+	−	+	+	−	−	−	+	−	−	−	−	+	−	−	2
Rhinoscepsis	+	−	−	−	−	+	−	+	+	−	−	+	−	+	−	−	+	+	−	−	−	−	−	−	−	+	−	−	0
Eutyphlus	−	−	−	−	−	+	−	−	+	+	+	+	−	+	−	−	+	−	+	−	−	−	−	−	+	+	−	−	2
Trisignis	−	−	−	−	+	+	−	−	+	+	−	+	−	+	−	−	+	−	+	−	−	−	−	−	−	+	−	−	2
Pseudactium	−	−	−	−	−	−	−	−	+	−	−	+	−	+	−	−	+	−	+	−	−	−	−	+	−	+	−	−	2
Ramecia	−	−	−	−	−	−	−	−	+	−	−	+	−	+	−	−	+	−	+	−	−	−	−	−	−	+	−	−	2
Trigonoplectus	−	−	−	−	−	−	−	−	−	−	+	−	−	+	−	−	−	+	−	−	+	−	−	+	−	−	+	−	2
Trimioplectus	−	−	−	−	−	−	−	−	−	−	−	+	−	+	−	−	−	+	−	−	+	−	−	+	−	−	+	−	2
Pycnoplectus	−	−	−	−	−	+	−	+	+	−	−	+	−	+	−	−	−	−	−	−	−	+	−	−	+	−	−	+	4
Acolonia	−	−	−	−	−	−	−	−	+	−	−	+	−	+	−	−	−	−	−	−	−	+	−	−	+	−	−	+	2
Barroeuplectoides	−	+	−	−	−	+	−	−	−	−	−	+	−	+	−	+	+	−	−	+	−	−	+	−	−	−	−	+	2
Tuberoplectus	−	+	+	−	−	+	−	−	−	−	−	+	−	+	−	+	+	−	−	+	−	−	+	−	−	−	−	+	2
Oropus*	−	±	−	+	+	+	−	−	+	−	−	+	−	+	−	−	+	−	+	−	+	−	−	−	−	−	+	−	4
Rhexidius*	−	+	−	+	+	+	−	−	+	−	−	+	−	+	−	−	+	−	+	−	+	−	−	−	−	−	+	−	4
Fletcherexius	−	+	−	−	+	+	−	−	+	−	−	+	−	+	−	−	+	−	+	−	+	−	−	−	−	−	+	−	4
Hexirhexius	−	+	−	+	+	+	−	−	+	−	−	+	−	+	−	−	+	−	+	−	+	−	−	+	−	−	+	−	4
Euboarhexius	−	+	−	+	+	+	−	−	+	−	−	+	−	+	−	−	−	−	+	−	+	−	−	−	−	−	+	−	4
Rhexius	−	+	−	−	−	+	−	−	+	−	−	+	−	+	−	−	−	−	+	−	+	−	−	+	−	−	+	−	4
Eurhexius	−	−	−	+	+	+	−	−	−	−	−	+	−	+	−	−	+	−	+	−	+	−	−	−	−	−	+	−	2
Euplectus	−	+	−	+	+	+	−	−	+	−	+	+	−	+	−	−	−	−	−	−	−	+	−	−	+	−	−	+	2
Rhexinia	−	−	−	+	−	+	−	−	+	−	−	+	−	+	−	−	−	−	+	−	+	−	−	+	−	−	+	−	2
Rhexiola	−	−	−	+	−	−	−	−	+	−	−	+	−	+	−	+	−	−	−	+	+	−	−	−	−	−	+	−	0
Hatchia	−	−	−	+	+	+	−	−	+	+	+	+	−	−	+	−	+	−	−	−	−	−	−	−	−	+	−	−	2
Bibloporus	−	+	−	+	−	+	+	−	+	+	−	+	−	+	−	−	+	−	+	−	−	−	−	−	+	+	−	−	2
Aboeurhexius	−	−	−	+	+	+	−	−	+	−	−	−	−	−	−	−	−	+	+	−	+	−	−	+	−	−	+	+	2

*Consult text for additional characters.

Rhinoscepsis

LeConte, 1878:382

(Plate 2, figs. 1–8)

Type-species: *Rhinoscepsis bistriatus* LeConte, 1878. Proc. Am. Philos. Soc. 17:382. Holotype ♀, no. 6174, Florida (MCZ). Monobasic.

Distribution: 12 species in this genus are found in South, Central, and North America (Florida).

Diagnosis: This genus has no close relatives among the genera considered in this study. The mesosternal area (2;1) is rather generalized by having nearly a full complement of relatively similar-sized and widely spaced foveae. It is unique in that it possesses a single median and two lateral metasternal foveae (2;1), and an unusual surface pattern on the mesosternum. The highly derived characters such as a frontal rostrum (2;2) and reduced eyes in both sexes easily set it apart from other genera of Euplectini.

Eutyphlus

LeConte, 1880:185

(Plate 3, figs. 1–8)

Type-species: *Eutyphlus similis* LeConte, 1880. Trans. Am. Entomol. Soc. 8:185–186. Holotype ♀, no. 6186, Virginia (MCZ). Monobasic.

Distribution: The 3 species in this genus were collected from Virginia, District of Columbia, and Pennsylvania.

Diagnosis: *Eutyphlus* has a fairly generalized structure and is not closely related to any particular genus. It has a median carina on the prosternum (3;2), as do *Thesium*, *Trisignis*, *Hatchia*, and *Bibloporus*, and possesses anterior prosternal foveae (3;2) like *Hatchia* and *Euplectus*. The latter two genera and *Bibloporus* differ by having foveae near the base of the antennae. *Thesium* has promesocoxal foveae (1;2), and *Trisignis* has modified median mesosternal foveae (4;5), characters which separate them from *Eutyphlus*.

Trisignis

Park and Schuster, 1955:1

(Plate 4, figs. 1–9)

Type-species: *Trisignis marshi* Park and Schuster, 1955. Natur. Hist. Misc. no. 148:2–4. Holotype ♂, near Freshwater, Humboldt Co., Calif. (CAS). Original designation.

Distribution: The genus contains 2 species from northwest California.

Diagnosis: This genus shows many generalized characteristics, but stands by itself in having one of the lateral mesosternal foveae and the median mesosternal foveae (4;5) asymmetrical in position from the normal. In addition to being directed to one side, the median mesosternal foveae have coalesced and appear as an extra-wide, single-median mesosternal fovea. Careful examination of a slide preparation will show the juncture of the foveal rings of the paired foveae along the midline—thus its placement in Group A rather than C. The 11th antennal segment (4;4) of *Trisignis* is somewhat longer than that of most genera, as is segment XI of *Rhinoscepsis* (2;3), but differs from that of *Rhinoscepsis* by having a subapical row of specialized setae.

Pseudactium

Casey, 1908:271

(Plate 5, figs. 1–7)

Type-species: *Pseudactium carolinae* Casey, 1908. Canadian Entomol. 40:271–272. Holotype ♀, no. 38627, North Carolina (USNM). Original designation.

Distribution: This genus contains 2 species from Pennsylvania in addition to the type of the genus.

Diagnosis: *Pseudactium* is placed (Diagram 2) with *Ramecia*, *Trigonoplectus*, *Trimioplectus*, and *Acolonia*, which are all somewhat more advanced than previous genera considered by not having a median pronotal impression. They also lack basal impressions on the pronotum and a median carina on the prosternum. All five genera have two foveae on sternite II. *Pseudactium* is very close in general appearance to *Ramecia*; slight differences exist in the surface pattern of the mesosternum (5;6), and the median mesosternal foveae of *Pseudactium* are more separated than the same foveae of *Ramecia* (6;5). *Pseudactium* also possesses weak basal carinae on tergite I, but *Ramecia* does not have them.

Ramecia
Casey, 1893:450
(Plate 6, figs. 1–9)

Type-species: *Euplectus arcuatus* LeConte, 1850. Boston Jour. Natur. Hist. 6:106–107. Holotype ♂, no. 6194, Athens, Ga. (MCZ). Designated by Bowman, 1934:144.

Distribution: The type-species of the genus is known from the southeastern United States, and the remaining two species in the genus are found respectively in Pennsylvania and Virginia.

Diagnosis: *Ramecia* is slightly more advanced than the closely related *Pseudactium* (see above) by having the median mesosternal foveae beginning to join at the base and by the absence of basal carinae on tergite I.

Trigonoplectus
Bowman, 1934:37–38
(Plate 7, figs. 1–8)

Type-species: *Trigonoplectus minutus* Bowman, 1934. The Pselaphidae of North America (Pittsburgh, Pa.): 38. Holotype ♂, near Clinton, Allegheny Co., Pa. (CM). Original designation.

Distribution: 2 species in the genus are both from Pennsylvania.

Diagnosis: The genera *Trigonoplectus* and *Trimioplectus* have a number of characters in common, such as the absence of metasternal foveae (7;4); no subhumeral foveae of the elytra (7;2); setate basal depressions and longitudinal carinae on tergite I (7;6); pit-like sensory area on the profemora (7;7); and somewhat similar genitalia (7;8). *Trigonoplectus* displays the more primitive condition of nearly closed mesocoxal cavities (7;4), but also the more advanced state of basal fusion of the median mesosternal foveae and reduced procoxal foveae (7;1).

Trimioplectus
Brendel, 1890:50
(Plate 8, figs. 1–8)

Type-species: *Trimioplectus obsoletus* Brendel, 1890. *In* Brendel and Wickham. Bull. Lab. Natur. Hist., State Univ. Iowa, 2:50–51. Lectotype ♂, no. 8290, Cedar Rapids, Iowa (ANSP). Monobasic.

Distribution: The genus consists of 2 species from the central United States.

Diagnosis: The relationships of *Trimioplectus* and *Trigonoplectus* are discussed under the latter genus (above). The median mesosternal foveae (8;5) are separated more in *Trimioplectus*; the mesocoxal cavities are more widely separated; and there is a reduction from two antebasal foveae on each elytron of *Trigonoplectus* (7;2) to one on the elytron of *Trimioplectus* (8;2).

Pycnoplectus
Casey 1897:552
(Plate 9, figs. 1–8)

Type-species: *Euplectus linearis* LeConte, 1850. Boston Jour. Natur. Hist. 6:104. Holotype ♂, no. 6188, Louisiana (MCZ). Designated by Wagner, 1975:158.

Distribution: 11 species occur in the United States from Minnesota to Texas and eastward to the Atlantic seaboard.

Diagnosis: *Pycnoplectus* is without metasternal foveae (9;1), as are *Acolonia* and the previous two genera, *Trigonoplectus* and *Trimioplectus*. It is also similar to *Acolonia* in that tergites I and II (9;5) have basal carinae and setae basal depressions, and that tergite IV is the largest. *Pycnoplectus* differs from all of the above-mentioned genera by possessing impressions on the pronotum and four foveae on sternite II (9;6). This genus has both primitive and advanced characteristics and is placed with those genera without foveae associated with the antennae, but it also has characteristics in common with those genera with foveae near the antennae (e.g., *Euplectus*).

Acolonia
Casey, 1893:453–454
(Plate 10, figs. 1–8)

Type-species: *Euplectus cavicollis* LeConte, 1878. Proc. Am. Philos. Soc. 17:387. Holotype ♂, no. 6192, Tampa, Fla. (MCZ). Monobasic.

Distribution: This monotypic genus is known only from Florida.

Diagnosis: The absence of impressions on the pronotum; no foveae on the metasternum (10;5); somewhat reduced foveae on the mesosternum (10;5); and open coxal cavities place *Acolonia* in a relatively advanced position in Group A. It differs from *Pycnoplectus* by having only two foveae on sternite II. *Acolonia* shows some similarities to *Ramecia*, *Trigonoplectus*, *Pseudactium*, and *Trimioplectus*, but the absence of metasternal foveae and the larger tergite IV easily separate it from these genera.

Barroeuplectoides
Park, 1942:101
(Plate 11, figs. 1–8)

Type-species: *Barroeuplectoides zeteki* Park, 1942. Northwestern Univ. Stud. Biol. Med. no. 1:101–104. Holotype ♀, Barro Colorado Island, Gatun Lake, Canal Zone (FMNH). Monobasic.

Distribution: *Barroeuplectoides* is known only from the type-specimen collected in the Canal Zone.

Diagnosis: This genus begins the series of Group A that have foveae associated with the lateral, dorsal, or both regions of the antennal tubercules. *Barroeuplectoides* is very closely related to *Tuberoplectus*. They differ in that the paired dorsal antennal foveae of *Barroeuplectoides* have separate openings (11;1), whereas those of *Tuberoplectus* have a common opening (12;1).

Barroeuplectoides and *Tuberoplectus* are placed by themselves and are branched off very low in Group A because they have a rather generalized mesosternal area (11;6) and promesocoxal foveae, as does *Thesium*. However, some of the foveae of the mesosternal area are reduced, and they also show a specialized condition of having the tarsal segments reduced from three to two (11;7). This character and a number of others of *Barroeuplectoides* and *Tuberoplectus* are shared with species in the Pyxidicerini.

Tuberoplectus
Park, 1952:96–97
(Plate 12, figs. 1–8)

Type-species: *Tuberoplectus plaumanni* Park, 1952. Chicago Acad. Sci. Spec. Pub. No. 9, Part 2:97–98. Holotype ♂, Nova Teutonia, Santa Catarina, Brazil (FMNH). Original designation.

Distribution: This genus is composed of the type species from Brazil and a second species from Mexico

Diagnosis: The relationship of *Tuberoplectus* to *Barroeuplectoides* is discussed under the latter (above). Other subtle differences include a more prominent median longitudinal depression on the pronotal disk of *Tuberoplectus* (12;2), and differences of the reticulations and punctation of the mesosternal areas (11;6 and

12;7). *Tuberoplectus* may be considered somewhat more advanced than *Barroeuplectoides* because of the common opening of the paired dorsal foveae near the antennae (12;1).

Oropus

Casey, 1886:196

(Plate 13, figs. 1–12; Plate 75, fig. 1; Plate 76, figs. 1–2; Plate 79, figs. 2, 4)

Type-species: *Oropus convexus* Casey, 1886. Bull. Calif. Acad. Sci. 2:198. Holotype ♂, Sonoma Co., Calif. (USNM). Original designation.

Distribution: 26 species of *Oropus* are found on the Pacific slope of North America.

Diagnosis: Diagram 3 shows *Oropus* with a cluster of 6 genera which all have four foveae on sternite II; tergite I (13;3) as the largest segment; a setate basal depression on tergite I (13;8); a subhumeral fovea on the elytron; a median discal impression on the pronotum (13;11); and several other common characters. Within this cluster, *Oropus* is closest to *Rhexidius*. Members of both genera may have foveae opening laterally and dorsally near the base of the antenna. However, the absence of dorsal foveae is the more frequent condition for species of *Oropus* (13;1). A prominent tooth is present on the lateral margin of the pronotum of species of *Oropus* (13;3), whereas this margin is crenulate on species of *Rhexidius*.

Rhexidius

Casey, 1887:478

(Plate 14, figs. 1–10)

Type-species: *Rhexidius granulosus* Casey, 1887. Bull. Calif. Acad. Sci. 2:478–479. Holotype ♂, Alameda Co., Calif. (USNM). Monobasic.

Distribution: 9 species have been described in this genus from the Coast Range of California, and 2 from Mexico.

Diagnosis: *Rhexidius* is considered somewhat more advanced than *Oropus*, with the median mesosternal foveae (13;2 and 14;3) showing some size reduction. *Rhexidius*, *Euboarhexius*, *Hexirhexius*, and some *Oropus* have foveae associated with the base of the antenna (14;1) both laterally and dorsally. *Rhexidius* does not have basal carinae on tergite I (14;6) as does *Hexirhexius*. *Euboarhexius* is unique in this association by not having metasternal foveae. *Rhexidius* has the surface of the pronotum with tubercules, but these do not have setae arising from them as does *Hexirhexius*. *Oropus* is distinguished by a single lateral tooth or spine on the pronotum near each lateral fovea.

Fletcherexius

Park, 1942:72–73

(Plate 15, figs. 1–7)

Type-species: *Eurhexius macrodactylus* Fletcher, 1928. Ann. Entomol. Soc. America. 21:203–204. Holotype ♀, no. 16806, S.W. Mexico City, Desierto de los Leons, 3,000 m., Mexico (CU). Monobasic.

Distribution: This monotypic genus is known only by three specimens from the type locality.

Diagnosis: *Fletcherexius* has foveae opening dorsally near the base of the antennae, but laterally opening foveae are absent. This condition is shared by *Rhexius*, but *Fletcherexius* is readily separated from the more advanced *Rhexius* by having metasternal foveae and closed mesocoxal cavities (15;2).

Hexirhexius, new genus

(Plate 16, figs. 1–7)

Type-species: *Euplectus canaliculatus* LeConte, 1850. Boston Jour. Natur. Hist. 6:107. Holotype ♂, no. 196 Columbia, S.C. (MCZ).
Rhexidius canaliculatus (LeConte). Casey, 1887:478.

Distribution: This genus consists of only 1 species, *Hexirhexius canaliculatus*, which is found east of the Mississippi River in the United States.

Diagnosis: Foveation of head (16;1) includes a pair of vertexal and gular foveae, the latter with single exterior opening; a pair of small foveae opening dorsally between base of antennae and vertexal foveae; and fovea opening laterally near base of antenna. Venter of head with normal acuminate setae and longitudinal median carina. Antenna clavate, last three segments (16;2) bearing enlarged tubular setae. Pronotum (16;4) with longitudinal median depression and biarcuate depression between lateral foveae, without basal depressions. Elytron (16;4) with sutural, two discal, and subhumeral foveae; sutural and subhumeral striae present. Prosternum with procoxal foveae, without median carina or anterior prosternal foveae. Mesosternum (16;3) with a pair of lateral mesosternal foveae, a pair of median mesosternal foveae (bases touching medianly), and a pair of lateral mesocoxal foveae. Without promesocoxal foveae; mesocoxal cavities closed. Metasternal foveae (16;3) present, metacoxae contiguous. Profemur with linear setate area, tarsi with primary and accessory claw (16;5). Setate basal depression and abdominal carinae on tergite I only, tergite I largest. Sternite II with two setate depressions, each bordered by two foveae.

Hexirhexius has many characters in common with *Oropus*, *Rhexidius*, and *Fletcherexius*. Only *Rhexidius*, *Euboarhexius*, and *Hexirhexius*, and some *Oropus* have two pairs of foveae associated with the base of the antennae, one pair opening dorsally and the other pair opening laterally. *Hexirhexius* differs from *Rhexidius* by the following: a microsetigerous area between the antennal tubercules of the male (16;1); pronotum with setae arising on tubercules; median mesosternal foveae closer together; a more complex male genitalia (16;7 and 14;8); and with basal carinae on tergite I. *Hexirhexius* has metasternal foveae, but *Euboarhexius* does not.

Euboarhexius
Grigarick and Schuster, 1966:31–33
(Plate 17, figs. 1–5)

Type-species: *Euboarhexius sinus* Grigarick and Schuster, 1966. Pan Pacific Entomol. 42:31–33. Holotype ♂, no. 111, Little River, Mendocino Co., Calif. (UCD). Monobasic.

Distribution: *Euboarhexius* was thought to be a monotypic genus from California, but we have seen specimens from Pennsylvania and Maryland, previously identified as *Rhexidius trogasteroides* Brendel, 1892, that are a second species of *Euboarhexius*.

Diagnosis: *Euboarhexius* has the same type of dorsal and lateral foveae near the base of the antenna as do *Rhexidius* (14;1) and *Hexirhexius* (16;1), but differs from these genera by not having metasternal foveae (17;2). Metasternal foveae are also absent from *Rhexius*, but *Euboarhexius* differs from *Rhexius* by having a median carina on the venter of the head; differences in the antennae (17;1); and absence of teeth on the lateral margin of the pronotum.

Rhexius
LeConte, 1850:102–103
(Plate 18, figs. 1–7; Plate 75, fig. 3; Plate 77, fig. 1)

Type-species: *Rhexius insculptus* LeConte, 1850. Boston Jour. Natur. Hist. 6:103. Holotype ♂, no. 6175, New Orleans, La. (MCZ). Monobasic.

Distribution: 18 species have been described in this genus from South, Central, and North America.

Diagnosis: The highly constricted anterior pronotum (18;2) and elongated segment I of the geniculate antennae clearly distinguish this genus from others. It represents the most advanced of the cluster of genera with four foveae on abdominal sternite II, because of the absence of metasternal foveae (18;1) and the lack of a ventral median carina on the head. Other genera in which antennal segment I is elongate have only two foveae on sternite II, and the antennae are not truly geniculate.

Eurhexius
Sharp, 1887:41
(Plate 19, figs. 1–8)

Type-species: *Eurhexius vestitus* Sharp, 1887. Biologia Centrali-Americana Coleoptera. Vol. 2, Part 1:41. Holotype ♂, Volcan de Chirqui 2,000–3,000 ft. Panama (BMNH). Subsequent designation.

Distribution: This neotropical genus contains about 25 species that are found from Uruguay to Guatemala.

Diagnosis: *Eurhexius* is a rather large genus and is grouped with 4 genera (Diagram 3) which have fewer foveae on sternite II than the previous genera with antennal fovea, and have procoxal foveae on the prosternum. It shares the presence of a ventral median carina on the head with *Euplectus*, but differs from it by having metasternal foveae (19;7). *Eurhexius* is readily distinguished by the pronotum (19;3), which is constricted anteriorly and has six prominent teeth on each side.

The type of the genus was not seen, and the generic characteristics are based on the type of *Eurhexius tropicus* Park, 1952.

Euplectus
Leach, 1817:80
(Plate 20, figs. 1–6; Plate 75, fig. 4)

Type-species: *Pselaphus nanus* Reichenbach, 1816. Monographia Pselaphorum. Lipsiae: 69. Europe. Monobasic.

Distribution: Members of this genus are widely distributed and it is one of the largest in the family, with slightly more than 100 species. Wagner (1975) recognized 10 species from North America. Approximately 7 are reported from the neotropical region.

Diagnosis: *Euplectus* is placed with *Eurhexius* because of the characters previously discussed under the latter (above). It is well separated from other genera by the presence of foveae opening near the base of the antennae laterally and dorsally; anterior prosternal foveae (20;1); setate basal depressions (20;2) on tergites I and II; and tergite I being the longest.

The generic characteristics are based on *Euplectus confluens* LeConte, 1850, as the type species of the genus was not seen.

Rhexinia
Raffray, 1890:103, 106
(Plate 21, figs. 1–8)

Type-species: *Rhexinia angulata* Raffray, 1890. Rev. d'Entomol. 9:196. New Granada (Colombia). Monobasic.

Distribution: The genus contains 3 species. The type species of the genus is found in Colombia; Tabasco, Colima, and Hustusco, Mexico; another species from Guadeloupe, Leeward Islands; and a third from Tucuman, Argentina.

Diagnosis: *Rhexinia* has a number of characters in common with *Rhexiola*, but differs by having the anterior pronotum constricted (21;3); a median discal impression on the pronotum; a setate basal depression on tergite I; and two foveae on sternite II. The generic characters are based on identified specimens of the Bonet Collection, and the type and paratype of the second and third species in the genus.

Rhexiola, new status
Park, 1952:57
(Plate 22, figs. 1–5)

Type-species: *Rhexinia (Rhexiola) mexicana* Park, 1952. Chicago Acad. Sci. Spec. Pub. no. 9, Part 2:58–59. Holotype ♀, Huatusco, Veracruz, Mexico (FMNH). Monobasic.

Distribution: This genus contains 2 species from the same locality in Mexico.

Diagnosis: *Rhexiola* is considered the most advanced genus in this cluster (Diagram 3) because of the absence of a median impression on the disk of the pronotum (22;2); the forked condition of the median mesosternal foveae (22;1); and the absence of foveae on sternite II.

Hatchia
Park and Wagner, 1961:9
(Plate 23, figs. 1-7)

Type-species: *Hatchia carinata* Park and Wagner, 1961. Univ. Wash. Pub. Biol. 16:9. Holotype ♂, Baring, Wash. (CAS). Monobasic.

Distribution: This monotypic genus is known only from the type locality.

Diagnosis: *Hatchia* is placed separately because of a number of distinct characteristics. It has some characters in common with *Euplectus*, such as anterior prosternal foveae (23;2); a longitudinal impression on the pronotal disk (23;3); and two foveae on sternite II. However, *Hatchia* has a prosternal carina (23;2) and metasternal foveae (23;4), and the median mesosternal foveae are joined at the base. *Hatchia* is placed near *Bibloporus* because they both have a median carina on the prosternum, but *Hatchia* differs by a number of characters, including the absence of a median carina on the venter of the head.

Bibloporus
Thomson, 1861:225
(Plate 24, figs. 1-9)

Type-species: *Euplectus bicolor* Denny, 1825. Monographia Pselaphidarum et Scydmaenidarum Britanniae: 17. Europe. Monobasic.

Distribution: 7 species are known from Europe, and the single species *bicanalis* (Casey, 1884) from New York.

Diagnosis: The lateral longitudinal impressions of the pronotum (24;3) of *Bibloporus* are unique among the genera we have observed in the tribe. *Bibloporus* is placed near *Hatchia* because of a number of similarities, but *Bibloporus* does not have anterior prosternal foveae or forked median mesosternal foveae (24;1). Foveae open near the base of the antennae both dorsally and laterally on *Bibloporus*. The generic characters are based on identified specimens of *bicolor* from the Park Collection and the type of *bicanalis*.

Aboeurhexius
Park, 1952:62
(Plate 25, figs. 1-9)

Type-species: *Aboeurhexius crenulatus* Park, 1952. Chicago Acad. Sci. Spec. Pub. no. 9, Part 2:62-63. Holotype ♀, El Pujal, San Luis Potosi, Mexico (FMNH). Monobasic.

Distribution: Known only from the type specimen collected in Mexico.

Diagnosis: *Aboeurhexius* is distinguished from all other genera in Group A by the presence of only a single median metasternal fovea (25;5); the absence of median mesosternal foveae; and the absence of prosternal foveae. The lack of median mesosternal foveae would place the genus in Group C in the proposed phylogenetic arrangement, but the continued presence of the specialized guard setae where the openings of the median mesosternal foveae usually occur (25;5); the presence of lateral foveae near the base of the antennae (25;1); and the length of the first antennal segment (25;4) warrant retaining *Aboeurhexius* as an advanced genus of Group A.

GROUP B

Of the 12 genera assigned to this group, 7 are found only in the western United States. *Trichonyx* is from New York state and Europe (a probable introduction to the United States—Park, 1953b), and 4 genera are neotropical.

Genera are placed in Group B primarily on the basis of the forked state of the lateral mesosternal foveae, but they also have the following characteristics in common: a pair, or remnants of a pair, of median mesosternal foveae; lateral mesocoxal foveae; procoxal foveae; without anterior prosternal foveae; without lateral longitudinal sulci on the pronotum; without foveae associated with the base of the antennae; tergites without basal carinae; mesosternum without median carina; and tarsi with three segments.

The proposed phylogenetic arrangement of Group B is shown in Diagram 4.

A tabular key (Table 4) to the genera included in this group is based on 18 characteristics.

Morius
Casey, 1893:445–446
(Plate 26, figs. 1–9)

Type-species: *Morius occidens* Casey, 1893. Ann. New York Acad. Sci. 7:446–447. Holotype ♀, Santa Cruz, Calif. (USNM). Monobasic.

Distribution: This monotypic genus is found only in California.

Diagnosis: *Morius* is placed by itself and low in the phylogenetic arrangement of Diagram 4 because of the presence of all the primary foveae (26;2,3,5); their large size; and closed mesocoxal cavities (26;5). The antenna (26;4) does not appear to have specialized setae. It is distinguished from all other genera in this group by a frontal rostrum (26;3).

Bontomtes, new genus
(Plate 27, figs. 1–9)

Type-species: *Bontomtes riparie*, new species.

Distribution: This monotypic genus is known only from the type locality.

Diagnosis: Foveation of head limited to a pair of vertexal and gular foveae, the latter with single exterior opening. Venter of head with normal acuminate setae; without median carina. Antenna clavate, last three segments (27;5) bearing normal, somewhat enlarged tubular setae. Pronotum with longitudinal median impression, biarcuate depression between lateral foveae, and impressions on basal margin. Elytron (27;4) with one sutural, discal, and subhumeral foveae. Prosternum with procoxal foveae (27;6), without anterior prosternal foveae, without median longitudinal carina. Mesosternum (27;2) with forked lateral mesosternal

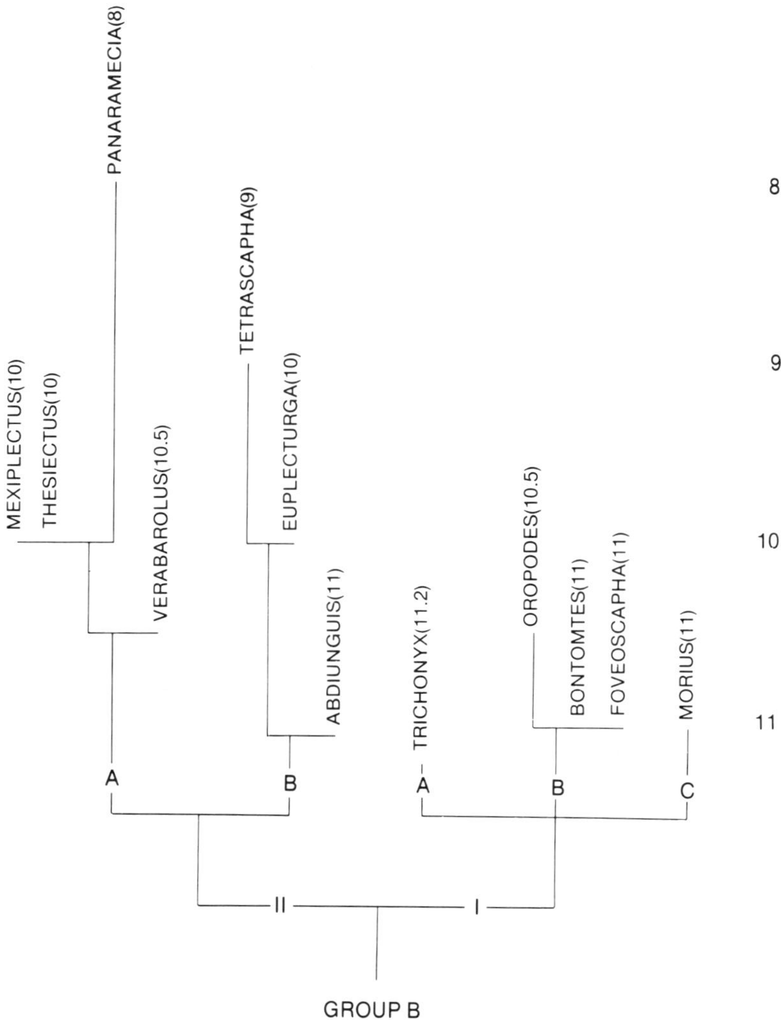

I. Sternite II with 4 foveae
   A. Pronotum with median impression; median mesosternal foveae forked; without rostrum
   B. Pronotum with median impression; median mesosternal foveae paired; without rostrum
   C. Pronotum without median impression; median mesosternal foveae paired; with rostrum
II. Sternite II with 2 foveae or none
   A. Pronotum without basal impressions
   B. Pronotum with basal impressions

DIAGRAM 4. Assignment of genera to Group B on the basis of a summation of the values of ancestral or derived characters listed in Tables 1, 2, and 4

TABLE 4. Tabular key to the genera of Group B

GENUS	HEAD			THORAX										TERGITES				S.
	Head with frontal rostrum	Head with ventral median carina	Median pronotal impression	Basal pronotal impressions	Elytron with subhumeral fovea	Elytron with antebasal foveae	Median carina on prosternum	Median mesosternal foveae paired	Median mesosternal foveae single, forked	Promesocoxal foveae	Metasternal foveae paired	Profemur with sensory modification	Metacoxae contiguous	Setate basal depression on tergite I only	Tergite length subequal	Tergite I longest	Tergites I and II longest	Foveae on sternite II
Morius	+	−	−	+	+	+	−	+	−	−	+	−	+	+	−	+	−	4
Bontomtes	−	−	+	+	+	+	−	+	−	−	+	−	+	−	+	−	−	4
Foveoscapha	−	−	+	+	+	+	−	+	−	−	+	−	+	−	−	+	−	4
Oropodes	−	−	+	+	+	+	−	+	−	−	−	−	+	+	+	−	−	4
Trichonyx	−	+	+	+	+	+	−	−	+	−	+	−	−	−	−	+	−	4
Abdiunguis	−	+	+	+	−	+	+	+	−	−	+	−	+	−	−	+	−	2
Euplecturga	−	−	+	+	+	+	−	+	−	+	−	−	+	−	+	−	−	2
Tetrascapha	−	−	−	+	−	+	−	+	−	−	−	−	+	+	−	+	−	2
Mexiplectus	−	−	−	−	+	+	+	+	−	−	+	−	+	−	−	+	−	2
Verabarolus	−	−	+	−	+	+	+	+	−	−	+	−	+	−	−	+	−	2
Thesiectus	−	−	−	−	+	+	+	+	−	−	+	+	+	−	+	−	−	2
Panaramecia	−	−	−	−	−	−	−	+	−	−	+	+	+	−	−	−	+	0

foveae, a pair of median mesosternal foveae, and a pair of lateral mesocoxal foveae. Procoxal foveae absent, mesocoxal cavities nearly closed. Metasternal foveae (27;2) present, metacoxae contiguous. Profemur without obvious sensory modifications; tarsi (27;1) with primary and accessory claws. Abdomen without carinae on tergites; tergite I with slight basal interruption without setae (27;8); tergites subequal in length. Sternite II with two setate depressions, each bordered with two foveae.

Bontomtes is associated with 4 genera in Group B that have four foveae on sternite II. It has the most characteristics in common with *Foveoscapha* and *Oropodes*. *Bontomtes* differs from *Oropodes* by having metasternal foveae (27;2), and from *Foveoscapha* by having tergites I to IV subequal. Tergite I of *Foveoscapha* is the longest.

Bontomtes riparie, new species

(Plate 27, figs. 1–9)

Male (Holotype): Red-brown; antennal segment XI, maxillary palpi yellow. Length 2.15 mm., width 0.5 mm. Head 480μ long, 450μ wide; vertexal foveae 165μ between centers, without guard setae. Ventral surface without specialized setae; gular foveae nude. Each eye with about 40 facets. Antenna 525μ long; segment I 10μ long x 75μ wide; II 85μ x 65μ; III–VIII 390μ long x 57μ wide; IX 75μ x 80μ, slightly asymmetrical; X 75μ x 90μ, symmetrical; XI 165μ x 105μ, apical one-third a cone (27;5); base of cone with flattened setae. Segment IV of maxillary palp with a row of specialized setae.

Pronotum with disk medially sulcate; pair of short carinae on each side of middle of base, another pair at each basolateral margin. Pronotum 510μ long, 535μ wide; elytron 705μ long, 450μ wide, with one discal, sutural, and subhumeral foveae (27;4). Winged.

Protrochanter with large, blunt spur (27;3); profemur 135μ wide; protibia with small conical spur on median of ventral face; mesofemur 140μ; metafemur 150μ wide; metatibia with curved apical spine 55μ long. Prosternum (27;6) slightly tumid on mid-line; mesosternum (27;2) without procoxal foveae; metasternum unmodified.

First visible tergite 555μ wide at base, with a pair of basolateral foveae (27;8); basal interruption 190μ wide, without setae. Sternite II with four foveae; III with median, transverse, hyaline plate (27;7), base 140μ wide, modified setae anterior and posterior to plate (27;7). Penial plate oblique, 90μ in width, longer than wide. Genitalia (27;9) 375μ long, 345μ deep.

Female: Unknown.

Distribution: The holotype male was collected at Tombstone Prairie (East of Santiam Pass) Linn Co., Oregon, under moss on a log, August 20, 1961, by W. Suter. The type (no. 879) is deposited at Davis (UCD).

Foveoscapha
Park and Wagner, 1961:13
(Plate 28, figs. 1-9)

Type-species: *Foveoscapha terracola* Park and Wagner, 1961. Univ. Wash. Pub. Biol. 16:13. Holotype ♂, Olympic Hot Springs, Olympic National Park, Wash. (FMNH). Monobasic.

Distribution: This monotypic genus is known from Washington and Oregon.

Diagnosis: The presence of metasternal foveae (28;6) separates *Foveoscapha* from *Oropodes*. *Foveoscapha* differs from the more closely related *Bontomtes* by having the metasternal foveae closer together and larger, and tergite I longest, compared to the subequal tergites of *Bontomtes*.

Oropodes
Casey, 1893:453
(Plate 29, figs. 1-10)

Type-species: *Oropodes orbiceps* Casey, 1893. Ann. New York Acad. Sci. 7:453-454. Holotype ♀, no. 38611, Los Angeles Co., Calif. (USNM). Monobasic.

Distribution: This genus contains 4 species from California and 1 from Oregon.

Diagnosis: *Oropodes* is considered more advanced than *Foveoscapha* and *Bontomtes* because of the absence of metasternal foveae (29;3). It also differs from these two genera by a setate basal depression on tergite I.

Trichonyx
Chaudoir, 1845:164
(Plate 30, figs. 1-8)

Type-species: *Pselaphus sulcicollis* Reichenbach, 1816. Monographia Pselaphorum: 62-63, Europe. Monobasic.

Distribution: This European species was reported by Park (1953b) from Flushing, Queens, N.Y., as a species that was probably introduced.

Diagnosis: *Trichonyx* is separated from the other genera with four foveae on sternite II (30;8) by the following characters: a ventral median carina on the head; metacoxae well separated (30;7); and the median mesosternal foveae joined to appear forked (30;4).

Abdiunguis
Park and Wagner, 1961:10
(Plate 31, figs. 1-8)

Type-species: *Abdiunguis fenderi* Park and Wagner, 1961. Univ. Wash. Pub. Biol. 16:10-12. Holotype ♂, Peavine Ridge near Minnville, Oregon (FMNH). Monobasic.

Distribution: This monotypic genus is found in Oregon and Washington.

Diagnosis: This genus is placed in a cluster (Diagram 4) with *Euplecturga* and *Tetrascapha* because of the common characters of two foveae on sternite II (32;2 and 33;4); basal depressions on the pronotum; and similarly shaped mesosternal shields (31;1, 32;7, and 33;1). *Abdiunguis* is separated from the other two genera by having metasternal foveae (31;1) and a median longitudinal carina on the venter of the head and the prosternum.

Euplecturga
Park and Wagner, 1961:15
(Plate 32, figs. 1-8)

Type-species: *Oropodes (Euplecturga) impressicollis* Park and Wagner, 1961. Univ. Wash. Pub. Biol. 16: 15-16. Holotype ♂, Forest Grove, Ore. (FMNH). Monobasic.
Euplecturga impressicollis Park and Wagner. Grigarick and Schuster, 1976.

Distribution: This genus is represented by 2 species from Oregon and a third from northern California.

Diagnosis: *Euplecturga* was considered a subgenus of *Oropodes*, but differs from it by having two foveae on sternite II (32;2), and prominent promesocoxal foveae (32;7). The promesocoxal foveae also separate it from the more closely related genera *Abdiunguis* and *Tetrascapha*.

Tetrascapha
Schuster and Marsh, 1957:149
(Plate 33, figs. 1-8; Plate 78, fig. 1)

Type-species: *Tetrascapha dasycerca* Schuster and Marsh, 1957. Pan-Pacific Entomol. 33:149-150. Holotype ♂, Mendocino, Mendocino Co., Calif. (CAS). Monobasic.

Distribution: The genus is limited to the type species from California and a second species from Oregon.

Diagnosis: The absence of a median discal impression on the pronotum (33;2) and the presence of setate basal depression on tergite I differentiate *Tetrascapha* from *Euplecturga* and *Abdiunguis*. Other prominent differences are discussed under those genera (above).

Mexiplectus
Park, 1943:174
(Plate 34, figs. 1-8)

Type-species: *Mexiplectus emersoni* Park, 1943. Bull. Chicago Acad. Sci. 7:174-175. Holotype ♂, Huichihuyan, San Luis, Veracruz, Mexico (FMNH). The author mistakenly listed the type as a female. Monobasic.

Distribution: This genus consists of 2 species from Veracruz and Chiapas, Mexico.

Diagnosis: *Mexiplectus* is in a cluster of 4 neotropical genera (Diagram 4) that are separated from other genera of Group B by the absence of basal impressions on the pronotum. *Verabarolus* is relatively close to *Mexiplectus*, as they both have tergite I (34;2 and 35;2) the longest, and the apex of antennal segment XI (34;4 and 35;3) bluntly rounded. The lateral mesocoxal and metasternal foveae of *Mexiplectus* 34;5) are larger than those of *Verabarolus* (35;4), which also differs by having a median discal impression on the pronotum (35;2).

Verabarolus
Park, 1942:104
(Plate 35, figs. 1–5)

Type-species: *Verabarolus subdendrus* Park, 1942. Northwestern Univ. Stud. Biol. Sci. Med. no. 1:104–106. Holotype ♀, Barro Colorado Island, Gatun Lake, Canal Zone, Panama (FMNH). Monobasic.

Distribution: This monotypic genus is known only from the type locality.

Diagnosis: This genus is most closely related to *Mexiplectus* for the reasons discussed under that genus (above). *Verabarolus* also has characters in common with *Thesiectus*, such as a median carina on the prosternum (36;1), and subhumeral foveae on the elytra; but the median discal impression on the pronotum (35;2) is a most obvious difference.

Thesiectus
Park, 1960:8
(Plate 36, figs. 1–6)

Type-species: *Thesiectus probus* Park, 1960. Trans. Am. Microscop. Soc. 79:9–10. Holotype ♂, 12 mi. south of Falmouth, Trelawny Parish, Cornwall County, Jamaica (FMNH). Monobasic.

Distribution: A monotypic genus recorded only from the type locality.

Diagnosis: *Thesiectus* shows characters in common with and different from *Verabarolus*, as discussed under that genus (above); however, it appears to be more closely related to *Panaramecia*. Antennal segment XI of *Thesiectus* (36;4) shows some elongation, and this segment is further elongated (37;3) in *Panaramecia*. Both of these genera have developed sensory modifications on the profemur. *Thesiectus* is unique within this cluster of genera by having the tergites subequal in length.

Panaramecia
Park, 1942:94
(Plate 37, figs. 1–7)

Type-species: *Panaramecia williamsi* Park, 1942. Northwestern Univ. Stud. Biol. Sci. Med. no. 1:94–96. Holotype ♂, Barro Colorado Island, Gatun Lake, Canal Zone, Panama (FMNH). Monobasic.

Distribution: This monotypic genus is known only from the type locality.

Diagnosis: *Panaramecia* is the most advanced of this cluster of genera, because of the absence of a median carina on the prosternum (37;1); no subhumeral foveae; reduced metasternal foveae (37;4); and the loss of foveae on sternite II. These properties, plus tergites I and II being the largest (37;2), readily characterize this genus.

GROUP C

The largest number of genera of Euplectini are assigned to this group, with 18 found in the United States and 25 in Central America and adjacent islands; 2 of these are cosmopolitan in distribution. Of the 36 genera placed in this group, 19 are monotypic.

The primary characteristic used to delegate genera to Group C is the presence of a single median mesosternal fovea, or the absence of a fovea in this position. Other characters in common include: lateral mesosternal foveae; the absence of promesocoxal foveae; head without frontal rostrum; without foveae near the base of the antennae; without lateral, longitudinal sulci on the pronotum; without basal depressions on the pronotum; without median carina on mesosternum; mesocoxal cavities open; and tarsi of three segments. The above description shows the lack of many characters in this group that were present in Groups A and B, and thus Group C is considered to be the most advanced group. Many of the genera assigned to Group C are closely related and may be clustered accordingly, but there appear to be several directions of evolution within the group as well as isolated genera still showing many of the characteristics of the previous two groups. Possible interpretations of these evolutionary trends are presented in Diagrams 5 and 6 in which Group C is divided into three parts.

Part I is limited to the genera *Mexigaster* and *Leptoplectus*. They are considered to be the most primitive genera of Group C, and are the only ones with four foveae on abdominal sternite II. The remaining genera, grouped on Diagram 5, have two foveae on abdominal sternite II and are placed in Part II. The genera of Part III are presented in Diagram 6, which includes those genera without foveae on sternite II. Part III contains the most advanced genera of Euplectini we have studied.

A tabular key which utilizes 23 characteristics for discrimination of genera of Group C is given in Table 5. (Those genera closely associated in the table and marked with an asterisk are not separable by the 23 characteristics used, and the text, figures, and bracket key should be consulted for characters that will distinguish them.)

Mexigaster
Park, 1952:110-111
(Plate 38, figs. 1-9)

Type-species: *Mexigaster boneti* Park, 1952. Chicago Acad. Sci. Spec. Pub. no. 9, Part 2:111-112. Holotype ♂, Nuevo Leon, Mexico (FMNH). Monobasic.

Distribution: This monotypic genus is known only from the type locality.

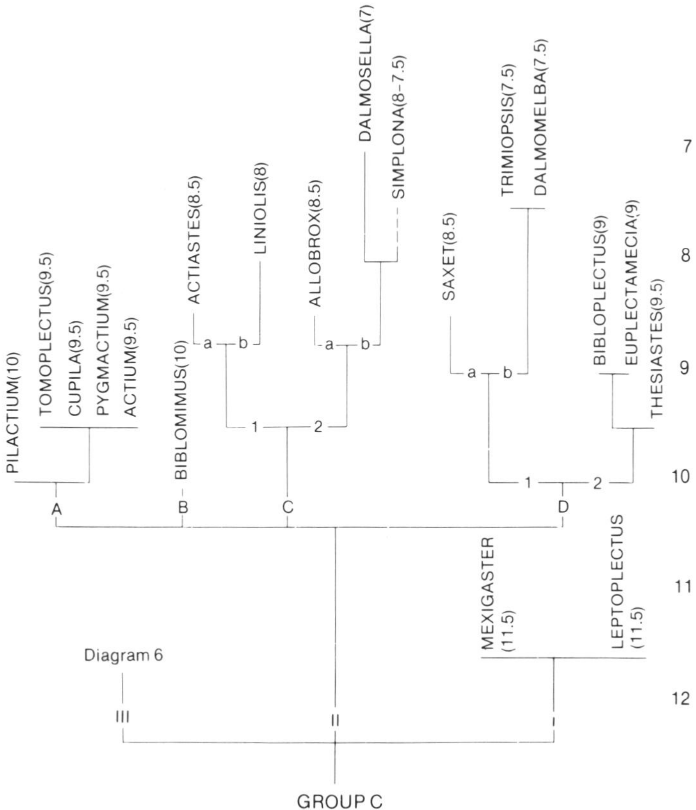

I. Sternite II with 4 foveae
   A. Venter of head and prosternum with median carina
   B. Venter of head and prosternum without median carina
II. Sternite II with 2 foveae
   A. Mesosternum with median fovea; prosternum without median carina;
      metasternal foveae paired
   B. Mesosternum with median fovea; prosternum with median carina;
      metasternal foveae paired
   C. Mesosternum with median fovea; prosternum without median carina;
      metasternal foveae single or absent
      1. Prosternum wtih procoxal foveae
         a. Elytron with subhumeral fovea
         b. Elytron without subhumeral fovea
      2. Prosternum without procoxal foveae
         a. Metasternum with one fovea
         b. Metasternum without foveae
   D. Mesosternum without median fovea; prosternum without medial carina;
      metasternal foveae paired
      1. Tergite I with basal carinae
         a. Prosternum with procoxal foveae
         b. Prosternum without procoxal foveae
      2. Tergite I without basal carinae

DIAGRAM 5. Assignment of genera to Group C, parts I and II, on the basis of a
summation of values of ancestral or derived characters listed in Tables 1, 2, and 5.

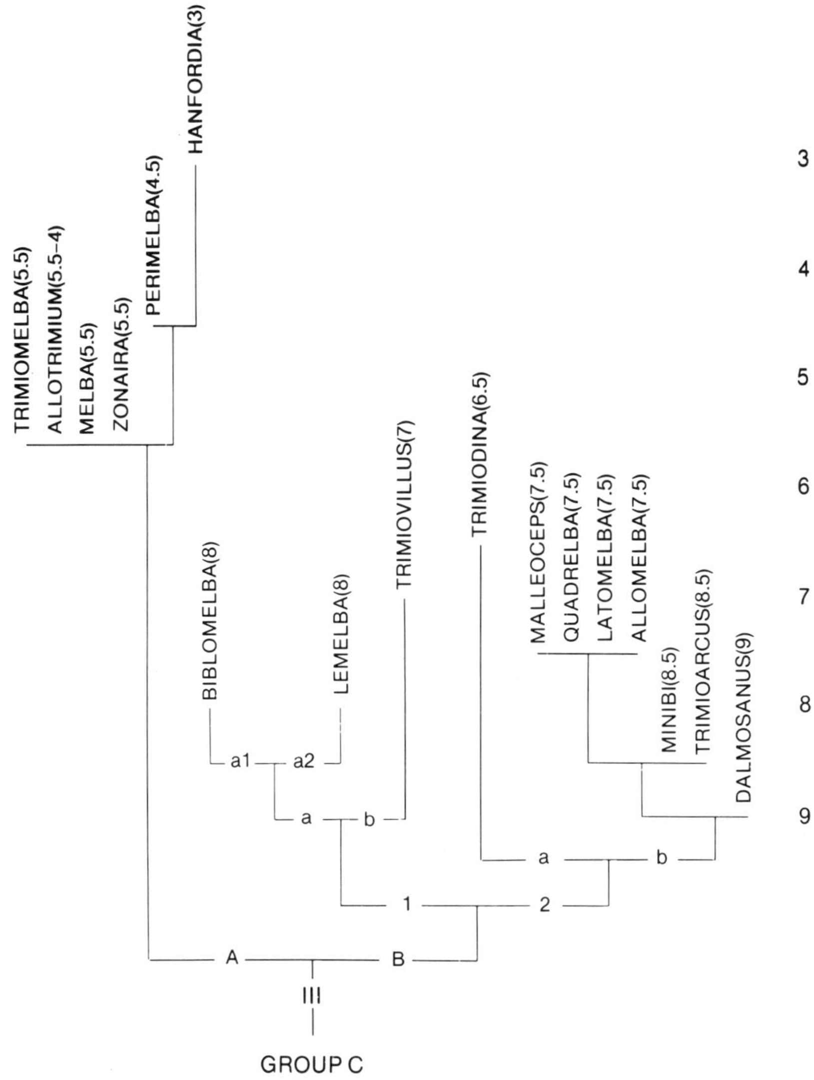

III. Sternite II without foveae
    A. Mesosternum without lateral mesocoxal foveae
    B. Mesosternum with lateral mesocoxal foveae
        1. Prosternum without procoxal foveae
            a. Pronotum without median impression
                a1. Prosternum with median carina
                a2. Prosternum without median carina
            b. Pronotum with median impression
        2. Prosternum with procoxal foveae
            a. Metasternal foveae absent
            b. Metasternal foveae paired

DIAGRAM 6. Assignment of genera to Group C, part III, on the basis of a summation of values of ancestral or derived characters listed in TAbles 1, 2, and 5

TABLE 5.　Tabular key to the genera of Group C

GENUS	H.	THORAX												TERGITES									S.
	Head with ventral median carina	Median pronotal impression	Elytron with subhumeral fovea	Elytra with antebasal foveae	Median carina on prosternum	Anterior prosternal foveae	Procoxal foveae	Median mesosternal fovea	Lateral mesocoxal foveae	Metasternal foveae paired	Metasternal fovea single	Profemur with sensory modification	Metacoxae contiguous	Setate basal depression on I only	Setate basal depression on I, II only	Setate basal depression on I, II, III	Basal carinae on I only	Basal carinae on I-II only	Basal carinae on I-III	Length subequal	I longest	IV longest	Foveae on sternite II
Part I.																							4
Mexigaster	+	+	+	+	+	−	+	+	+	+	−	+	+	−	−	−	+	−	−	−	+	−	4
Leptoplectus	−	+	+	+	−	+	+	+	+	+	−	−	+	−	+	−	−	−	−	−	−	+	4
Part II.																							2
Biblomimus	−	−	+	+	+	−	+	+	+	+	−	+	+	−	−	−	+	−	−	+	−	−	2
Liniolis	−	−	−	+	−	−	+	+	+	−	−	+	+	−	−	−	−	−	+	−	−	+	2
Tomoplectus	−	−	+	+	−	−	+	+	+	+	−	−	+	−	−	−	+	−	−	+	−	−	2
Cupila*	−	−	+	+	−	−	+	+	+	+	−	+	+	−	−	−	±	−	−	−	+	−	2
Pilactium	−	−	+	+	−	−	+	+	+	+	−	+	+	+	−	−	−	−	−	−	+	−	2
Pygmactium*	−	−	+	+	−	−	+	+	+	+	−	+	+	−	−	−	+	−	−	−	+	−	2
Actium	−	−	+	+	−	−	+	+	+	+	−	+	+	−	−	−	+	±	±	+	−	−	2
Euplectamecia	+	−	+	+	−	−	+	−	+	+	−	−	+	−	−	−	−	−	−	+	−	−	2
Thesiastes	−	+	+	+	−	−	+	−	+	+	−	+	+	−	−	+	−	−	−	+	−	−	2
Bibloplectus	−	−	+	+	−	−	+	−	+	+	−	+	+	+	−	−	−	−	−	+	−	−	2
Saxet	−	−	+	+	−	−	+	−	+	+	−	−	+	−	−	−	−	+	−	−	−	+	2
Dalmomelba	+	−	−	+	−	−	−	−	+	+	−	−	+	−	−	−	+	−	−	+	−	−	2
Trimiopsis	−	+	−	+	−	−	−	−	+	+	−	−	+	−	−	−	+	−	−	+	−	−	2
Allobrox	−	−	+	+	−	−	−	+	+	−	+	+	+	+	−	−	−	−	−	+	−	−	2
Actiastes	−	−	+	+	−	−	+	+	+	−	−	+	+	−	−	−	+	−	−	−	+	−	2
Simplona	−	−	+	+	−	−	−	+	+	−	−	+	+	±	−	−	+	−	−	−	+	−	2
Dalmosella	−	−	−	+	−	−	−	+	+	−	−	+	+	−	−	−	+	−	−	−	+	−	2
Part III.																							0
Biblomelba	−	−	−	+	+	−	−	+	+	+	−	+	−	−	−	−	+	−	−	−	+	−	0
Lemelba	−	−	+	+	−	−	−	+	+	+	−	+	−	−	−	−	+	−	−	−	+	−	0
Trimioarcus	−	−	−	+	−	−	+	+	+	+	−	+	+	−	−	−	+	−	−	−	+	−	0
Dalmosanus	−	−	+	+	−	−	+	+	+	+	−	+	+	−	−	−	−	−	−	−	+	−	0
Minibi	−	−	−	+	−	−	+	+	+	+	−	+	+	−	−	−	−	−	+	+	−	−	0
Allomelba	−	−	−	+	−	−	+	−	+	+	−	−	+	−	−	−	+	−	−	+	−	−	0
Latomelba	−	−	−	+	−	−	+	−	+	+	−	−	−	−	−	−	+	−	−	−	+	+	0
Trimiodina	−	−	−	+	−	−	+	−	+	−	−	+	+	−	−	−	−	−	−	−	+	−	0
Trimiovillus	−	+	−	+	−	−	−	−	+	+	−	+	+	−	−	−	+	−	−	−	+	−	0
Quadrelba	−	−	−	+	−	−	+	−	+	+	−	−	−	−	−	−	+	−	−	−	+	−	0
Malleoceps	−	−	−	+	−	−	+	−	+	+	−	−	+	−	−	−	+	−	−	−	+	−	0
Trimiomelba*	−	−	−	+	−	−	−	+	−	+	−	+	+	−	−	−	+	−	−	−	+	−	0
Allotrimium*	−	−	−	+	−	−	−	+	−	±	−	+	+	−	−	−	+	−	−	−	+	−	0
Melba*	−	−	−	+	−	−	−	+	−	+	−	+	+	−	−	−	+	−	−	−	+	−	0
Zonaira	−	−	−	+	−	−	−	+	−	+	−	+	+	−	−	−	−	−	+	−	+	−	0
Perimelba	−	−	−	+	−	−	−	−	−	+	−	+	+	−	−	−	−	−	+	+	−	−	0
Hanfordia	−	−	−	−	−	−	−	−	−	−	−	+	−	−	−	−	−	+	−	+	−	−	0

*Consult text for additional characters.

Diagnosis: *Mexigaster* is placed in Part I of Group C with *Leptoplectus* because of the presence of a single median mesosternal fovea (38;5 and 39;7) and four foveae on sternite II (38;7). It differs from *Leptoplectus* in many characters, some of the most obvious being the presence of longitudinal carinae on the venter of the head and prosternum (38;1) and the absence of anterior prosternal foveae on *Mexigaster*.

Leptoplectus

Casey, 1908:266–267

(Plate 39, figs. 1–8)

Type-species: *Euplectus pertenuis* Casey, 1884. Contributions to the Descriptive and Systematic Coleopterology of North America, Part II:109–110. Holotype ♀, no. 38607, Washington, D.C. (USNM). Designated by Bowman, 1934:144.

Distribution: *Leptoplectus pertenuis* is widely distributed in the eastern United States. Four species were described in the genus by Casey, but Wagner (1975) placed one of the species in *Euplectus* and synonymized two with *pertenuis*. Another species was reported by Raffray (1911) from New Guinea, but the disjunct distribution makes the generic designation somewhat questionable.

Diagnosis: The similarities of *Leptoplectus* to *Mexigaster* as members of Part I are discussed under the latter (above). Additional characters that distinguish *Leptoplectus* from *Mexigaster* include: a setate basal depression on tergites I and II (presence variable on II); tergite IV largest (39;2); and the prosternum with anterior prosternal foveae (39;4). Antennal segment XI (39;6) is rounded apically on *Leptoplectus*, but acuminate (38;3) on *Mexigaster*.

Biblomimus

Raffray 1903:545

(Plate 40, figs. 1–9)

Type-species: *Biblomimus minutus* Raffray, 1903. Ann. Soc. Entomol. France 72:545. Holotype ♀, St. Vincent, Windward Islands (MNHN). Monobasic.
Ramecia minutus (Raffray). Raffray, 1908. Ann. Soc. Entomol. France 77:36.
Biblomimus minutus Raffray. Park, 1942, Northwestern Univ. Stud. Biol. Med. no. 1:93.

Distribution: Known from 4 species, one each in Honduras, Leeward Islands, Windward Islands, and Mexico.

Diagnosis: *Biblomimus* is in the group of genera of Part II that have two foveae on sternite II. This genus has many characters in common with the adjacent cluster (Diagram 5) of 5 genera including *Actium*, but differs from all of them by having a longitudinal median carina on the prosternum (40;8). Generic characteristics were based on the type of *Biblomimus vertexalis* Park.

Liniolis, new genus

(Plate 41, figs. 1–7)

Type-species: *Euplectus crinitus* Brendel, 1890. Brendel and Wickham. Bull. Lab. Natur. Hist., Univ. Iowa 2:55–56. Lectotype ♂, no. 8278, Illinois (ANSP).
Ramecia crinita (Brendel). Casey, 1893. Ann. New York Acad. Sci. 7:451.

Distribution: *Liniolis crinitus* is found in north-central and northeastern United States.

Diagnosis: Foveation of head limited to a pair of vertexal and gular foveae, latter with common exterior opening; venter of head with simple acuminate setae; without median carina. Antennal club three-segmented, segments nearly symmetrical. Pronotum without longitudinal median or lateral depressions; with biarcuate depression between lateral foveae (41;3); basal depressions absent. Elytra with sutural and two discal foveae, without subhumeral foveae (41;3). Prosternum with procoxal foveae (41;1), without anterior prosternal foveae or longitudinal median carina. Mesosternum (41;4) with a pair of lateral mesosternal foveae, a single median mesosternal fovea; without promesocoxal foveae or metasternal foveae. Mesocoxal cavities open; profemur (41;2) with a row of sensory setae; metacoxae contiguous. Tergites I to III with basal carinae (41;6), without basal setate depression; tergite IV longest. Sternite II with a pair of lateral foveae.

Liniolis is distinguished from the genera *Euplectus* and *Ramecia* by having a single fovea, rather than paired, median mesosternal foveae. This genus is placed near *Actiastes* but shows the more advanced character of the absence of a subhumeral fovea of the elytron. The greater length of tergite IV also separates it from *Biblomimus* as well as the adjacent cluster of genera (Diagram 5) that contains *Actium*.

Tomoplectus
Raffray, 1898:266
(Plate 42, figs. 1–8)

Type-species: *Tomoplectus cordicollis* Raffray, 1898. Rev. d'Entomol. 17:267. Mexico. Monobasic.

Distribution: The 2 species in this genus are both found in Mexico.

Diagnosis: The genus *Tomoplectus* is placed in a cluster of closely related (Diagram 5) genera including *Cupila*, *Pilactium*, *Pygmactium*, and *Actium*. It differs from them by not having sensory setae on the profemur (42;6), and possessing a more conical antennal segment XI (42;3). The length of tergites I to IV are subequal as in *Actium*, but *Actium* has one or two antebasal foveae on the elytron, compared to only one on the elytron of *Tomoplectus*. The generic characters were based on the type of *Tomoplectus mexicanus* Park, 1952.

Cupila
Casey, 1897:561
(Plate 43, figs. 1–7

Type-species: *Trimium clavicornis* Mäklin, 1852. Bull. Soc. Imper. Naturalistes du Moscou 25:371–372. Oregon. Monobasic.

Distribution: 3 species are known, one each from Washington, Oregon, and northern California.

Diagnosis: *Cupila* is separated from *Actium* by having tergite I longest, and from *Pilactium* by not having a basal setate depression on tergite I. *Pygmactium* and *Cupila* cannot be separated by the characters given in Table 5. The shape of the mesosternal shield (43;4) of *Cupila* is wider and contains more setae than does the elongate mesosternal shield of *Pygmactium*, which always has basal carinae on sternite II; but the presence or absence of basal carinae on tergites I and II of *Cupila* is variable, depending on the species. *Cupila* tends to be dark red-brown to blackish, and *Pygmactium* is light brown. Generic characteristics are based on identified specimens of *Cupila clavicornis* and *C. excavata* Park and Wagner, 1961, and the type of *C. multifossa* Grigarick and Schuster, 1968.

Pilactium
Grigarick and Schuster, 1970:37
(Plate 44, figs. 1–9)

Type-species: *Pilactium summersi* Grigarick and Schuster, 1970. Pan-Pacific Entomol. 46:37. Holotype ♂, no. 880, Quincy, Plumas Co., Calif. (UCD). Monobasic.

Distribution: This monotypic genus is represented by the type-species in northern California and a second species from southern Oregon.

Diagnosis: *Pilactium* is distinguished from the 4 genera with which it is placed (Diagram 5) by having a prominent basal setate depression on tergite I (44;5).

Pygmactium
Schuster and Grigarick, 1968:112–113
(Plate 45, figs. 1–9; Plate 77, fig. 3; Plate 78, fig. 2)

Type-species: *Pygmactium steevesi* Schuster and Grigarick, 1968. Pan-Pacific Entomol. 44:113. Holotype ♂. Highlands Hammocks State Park, Florida (FMNH). Original designation.

Distribution: The type-species has been found only in Florida, and a second species occurs throughout Alabama, North Carolina, and Tennessee.

Diagnosis: *Pygmactium* is not separated from *Cupila* by the characters used in Table 5. As discussed under the latter genus (above), differences in the configuration of the mesosternal shields (43;4 and 45;4) and the presence of basal carinae on sternite II of *Pygmactium* will separate these genera. The wide basal carinae of tergite I (45;7) of *Pygmactium* is also quite distinctive for the genus.

Actium
Casey, 1886:201
(Plate 46, figs. 1–7)

Type-species: *Trimium californicum* LeConte, 1878. Proc. Am. Philos. Soc. 17:383. California (type depository unknown). Type by subsequent designation.

Distribution: Approximately 35 species are known from North America, most of which occur along the Pacific slope. One species is reported from Mexico, one from Puerto Rico, and another from Jamaica, but we have not seen them.

Diagnosis: *Actium* differs from 4 of the genera in the cluster (Diagram 5) in which it is included by having the tergites of subequal length. The tergites of *Tomoplectus* are also subequal in length, but the profemur of *Tomoplectus* is devoid of the specialized sensory setae that occur on *Actium*. Species of *Actium* may or may not have basal carinae on tergites I and II. The generic characteristics were based on identified material of *A. californicum* and the types of several species of *Actium*.

Euplectamecia
Park, 1952:102–103
(Plate 47, figs. 1–8)

Type-species: *Euplectamecia hidalgensis* Park, 1952. Chicago Acad. Sci. Spec. Pub. no. 9, Part 2:103. Holotype ♂. Chapulhuacán, Hidalgo, Mexico (FMNH). Monobasic.

Distribution: This monotypic genus is known only from the type locality.

Diagnosis: *Euplectamecia* is closely related to *Thesiastes* and *Bibloplectus* with respect to the configuration of the mesosternal shield and foveae of the mesosternum (47;4, 48;1, and 49;1). All three genera lack a median mesosternal fovea. The ventral median carina of the head; absence of sensory modifications on the profemur; and lack of basal setate depressions on the tergites (47;5) set *Euplectamecia* apart from the other two genera.

Thesiastes
Casey, 1893:457
(Plate 48, figs. 1–7)

Type-species: *Euplectus debilis* LeConte, 1878. Proc. Am. Philos. Soc. 17:386. Holotype ♂, no. 6197, Tampa, Fla. (MCZ). Designated by Bowman, 1934:144.

Distribution: About 15 species are distributed throughout North, Central, and South America, Europe, Africa, Asia, and Indonesia.

Diagnosis: This genus is relatively close to *Bibloplectus* in that both have basal setate depressions on at least tergite I (48;5 and 49;4), and the profemur has pit-like sensory modifications (48;6). *Thesiastes* differs from *Bibloplectus* by having a median discal impression on the pronotum (48;2), and basal setate depressions on tergites I to III—the depression on tergite III may be very weak on some specimens.

Bibloplectus
Reitter, 1881:529
(Plate 49, figs. 1-7)

Type-species: *Pselaphus ambiguus* Reichenbach, 1816. Monographia Pselaphorum: 67-68. Europe. Designated by Bowman, 1934:144.

Distribution: Europe, Asia, Africa, and North America. Of over 20 species, 4 are found in the eastern United States.

Diagnosis: *Bibloplectus* has many characters similar to *Thesiastes*, but lacks the median discal impression on the pronotum (49;2), and shows a reduced development of basal setate depressions on the tergites (49;4). Generic characteristics for this large genus were taken from a paratype of *Bibloplectus sobrinus* Casey, 1897.

Saxet, new genus
(Plate 50, figs. 1-9)

Type-species: *Euplectus decoris* Casey, 1884. Contributions to the Descriptive and Systematic Coleopterology of North America, Part 2:113-114. Holotype ♀, no. 38605, Columbus, Texas (USNM).
Ramecia decora (Casey). Casey, 1893. Coleopterological Notices, V. Ann. New York Acad. Sci. 9:451.

Distribution: This monotypic genus is known from the type locality and Arizona.

Diagnosis: Foveation of head limited to a pair of vertexal and gular foveae. Venter of head with normal acuminate setae; without median ventral carina. Antennae with three-segmented club (50;3). Pronotum without median discal impression or lateral depressions; with biarcuate depression between lateral foveae (50;2); basal depressions absent. Elytron with sutural, one antebasal, and subhumeral foveae (50;2). Prosternum with procoxal foveae; without anterior prosternal foveae or longitudinal median carina. Mesosternum (50;1) with a pair of lateral mesosternal foveae; without median mesosternal fovea; lateral mesocoxal foveae present; promesocoxal foveae absent. A pair of metasternal foveae present (50;1). Mesocoxal cavities open; metacoxae contiguous; profemur without prominent sensory modification; mesotarsus with primary and accessory claw (50;5). Tergites I to III with basal carinae (50;4) without basal setate depressions; tergite IV longest. Sternite II with a pair of lateral foveae.

Saxet is separated from *Euplectus* and *Ramecia* by not having median mesosternal foveae. The mesosternal area of *Saxet* shows similarities to the cluster of genera containing *Euplectamecia*, *Thesiastes*, and *Bibloplectus*. It differs from these by having basal carinae (50;4) and no basal setate depressions on the tergites, and by tergite IV being the longest. It is separated from *Trimiopsis* and *Dalmomelba* by having procoxal foveae.

Dalmomelba
Park, 1954a:6
(Plate 51, figs. 1-7)

Type-species: *Dalmomelba cazieri* Park, 1954a. American Mus. Novitates no. 1674:6-8. Holotype ♂, South Bimini Island, Bahamas, British West Indies (AMNH). Monobasic.

Distribution: A monotypic genus collected only from the type locality.

Diagnosis: *Dalmomelba* and *Trimiopsis* are related to the previous 4 genera discussed by not having a median mesosternal fovea (51;4 and 52;4) but are separated from them by the absence of procoxal foveae (51;1 and 52;1). *Dalmomelba* differs from *Trimiopsis* by having a ventral median carina on the head and by lacking a median discal impression on the pronotum.

Trimiopsis
Reitter, 1882a:149
(Plate 52, figs. 1–7)

Type-species: *Trimiopsis caviceps* Reitter, 1882a. Deutsche Entomol. Zeitschr. 26:150. Holotype ♀, La Luzera, Colombia. Monobasic.

Distribution: This genus consists of the type species and three species from Central America, one each from Panama, Guatemala, and Mexico.

Diagnosis: In addition to the similarities and differences between *Trimiopsis* and *Dalmomelba* as discussed under the latter (above), the epipleural stria is missing from the elytron of *Trimiopsis* (52;2) but is present on *Dalmomelba* (51;2). Tergites I to IV are subequal in length for *Trimiopsis*, but tergite I is considerably longer for *Dalmomelba*. Generic characteristics were based on the type of *Trimiopsis furcalis* Park, 1942.

Allobrox
Fletcher, 1928:205
(Plate 53, figs. 1–5; Plate 54, figs. 1–7; Plate 77, fig. 2)

Type-species: *Allobrox dampfi* Fletcher, 1928. Ann. Entomol. Soc. America 21:206. Holotype ♂, no. 805, SW Mexico City, Desierto de Los Leons, 3,000 m, Mexico (CU). Monobasic.

Distribution: 2 species are known from Mexico, and an undescribed species was found in Arizona.

Diagnosis: *Allobrox* is unique among the genera of Group C by having a single median metasternal fovea (53;1). This character, plus the loss of procoxal foveae, indicate an advanced condition.

Allobrox mexicana (Park), new status
(Plate 54, figs. 1–7)

Cupila (Cutrimia) mexicana Park, 1945. Bull. Chicago Acad. Sci. 7:336–337. Holotype ♂, nr. Mexico City, Mexico (FMNH).

This synonymy came to our attention following a dissection of the type.

Actiastes
Casey, 1897:561
(Plate 55, figs. 1–8)

Type-species: *Trimium foveicolle* LeConte, 1878. Proc. Am. Philos. Soc. 17:384. Holotype ♂, no. 6177, Massachusetts (MCZ). Monobasic.

Distribution: 7 species are limited to the eastern United States, but 1 species extends across the continent to British Columbia and another is restricted to southern California.

Diagnosis: *Actiastes* possesses a number of characteristics in common with the cluster of genera containing *Actium*, but differs from them by the absence of metasternal foveae.

Simplona
Casey, 1897:561
(Plate 56, figs. 1–9; Plate 78, fig. 3)

Type-species: *Simplona arizonica* Casey, 1897. Ann. New York Acad. Sci. 9:562. Holotype ♀, no. 38649, Arizona (USNM). Monobasic.

Distribution: This monotypic genus is found only in Arizona.

Diagnosis: *Simplona* and *Dalmosella* are placed in an advanced position of Part II of Group C (Diagram 5) because they have neither procoxal foveae (56;1 and 57;4) nor metasternal foveae (56;4 and 57;7). Antennal segment XI of *Simplona* is longer (56;3) than that of *Dalmosella* (57;2). They are also separated by the presence of a subhumeral fovea and epipleural stria for *Simplona* (56;2), structures which are absent from *Dalmosella* (57;3).

Dalmosella
Casey, 1897:570
(Plate 57, figs. 1-8)

Type-species: *Dalmosella tenuis* Casey, 1897. Ann. New York Acad. Sci. 9:570-571. Holotype ♀, no. 38655, Westmoreland Co., Pa. (USNM). Designated by Park, 1942. Northwestern Univ. Stud. Biol. Sci. Med. no. 1:115.
Melba tenuis (Casey). Raffray, 1898:237.
Dalmosella tenuis Casey. Park, 1942:116.

Distribution: *Dalmosella tenuis* is known from Pennsylvania and Illinois. Two species, *americanum* and *simplex* from the southern part of the United States, were also relegated to this genus by Casey, but this assignment has been questioned by Park and Raffray.

Diagnosis: Differences in the mesosternal shields (56;4 and 57;7) separate *Dalmosella* from *Simplona*, as well as do the antennal differences and reduced foveation of the elytron as is discussed for *Simplona* (above).

Biblomelba
Park, 1952:106
(Plate 58, figs. 1-6)

Type-species: *Biblomelba profunda* Park, 1952. Chicago Acad. Sci. Spec. Pub. no. 9, Part 2:106-107. Holotype ♂, Chichicaste, Tabasco, Mexico (FMNH). Monobasic.

Distribution: Known only from the type specimen collected in Mexico.

Diagnosis: *Biblomelba* is the first genus considered in Part III of Group C (Diagram 6), which includes those genera without foveae on sternite II. This genus has several characters in common with *Lemelba*, such as the absence of procoxal foveae (58;1), and metacoxae that are well separated (58;5 and 59;6). *Biblomelba* is distinguished from all genera of Part III by the presence of a median longitudinal carina on the prosternum (58;1).

Lemelba
Park, 1953a:1
(Plate 59, figs. 1-9)

Type-species: *Lemelba davisi* Park, 1953a. Natur. Hist. Misc. no. 121:2-4. Holotype ♂, Everglades, near Miami, Dade Co., Fla. (FMNH). Monobasic.

Distribution: This genus contains the type species from the Florida Everglades and a second species from Cuba.

Diagnosis: *Lemelba* has widely separated metacoxae (59;6), as do *Biblomelba*, *Latomelba*, and *Quadrelba*. It differs from these 3 genera by the elytron having a subhumeral fovea.

Trimioarcus
Park, 1952:130
(Plate 60, figs. 1-7)

Type-species: *Trimioarcus incisurus* Park, 1952. Chicago Acad. Sci. Special Pub. no. 9, Part 2:131-132. Holotype ♂. Monterrey, Nuevo Leon, Mexico (FMNH). Monobasic.

Distribution: This monotypic genus is known only from the type locality.

Diagnosis: *Trimioarcus* and *Dalmosanus* appear to be the most generalized of a cluster of 5 related genera (Diagram 6). Within this cluster, there is a trend toward reduction in size of the procoxal foveae, and loss of the median mesosternal fovea. *Trimioarcus* is most similar to *Dalmosanus*, but differs by the elytron not having a subhumeral fovea or epipleural stria (60;2), and by possessing basal abdominal carinae on tergite I.

Dalmosanus
Park, 1952:126–127
(Plate 61, figs. 1–6)

Type-species: *Dalmosanus inoculus* Park, 1952. Chicago Acad. Sci. Special Pub. no. 9, Part 2:127–128. Holotype ♀, Huatusco, Veracruz, Mexico (FMNH). Monobasic.
Triangusella acuta Park, 1952.132 l.c. Holotype ♂, Atoyac, Veracruz, Mexico (FMNH). Monobasic. New synonymy for the genus.

Distribution: This genus contains the 2 species listed above.

Diagnosis: *Dalmosanus* is related to *Trimioarcus* and also to *Minibi*. It differs from both of these genera by the presence of a subhumeral fovea on the elytron (61;3). *Triangusella* is synonymized with *Dalmosanus* because the type species of both have the same characters listed for *Dalmosanus* in Table 5. In addition, the mesosternal shields have the same pattern, and both holotypes have a pair of pores on tergites I–III.

Minibi, new genus
(Plate 62, figs. 1–6)

Type-species: *Melba* (*Melba*) *zonula* Park, 1954a. American Mus. Novitates no. 1674:13–15. Holotype ♂, South Bimini Island (AMNH).

Distribution: Known only from the type locality.

Diagnosis: Foveation of head limited to a pair of vertexal and a pair of gular foveae, the latter open separately. Venter of head without longitudinal median carina. Antenna with three-segmented club (62;3). Pronotum without median discal or lateral depressions; with biarcuate antebasal sulcus between vestigial lateral foveae; basal depressions absent. Elytra with sutural and one discal fovea; epipleural stria present, but subhumeral fovea absent (62;2). Prosternum with procoxal foveae (62;1); without anterior prosternal foveae or median longitudinal carina. Mesosternum (62;5) with a pair of lateral mesosternal foveae, a single median mesosternal fovea, and a pair of lateral mesocoxal foveae. Promesocoxal foveae absent; a pair of metasternal foveae present. Mesocoxal cavities open; profemur with sensory modification; metacoxae contiguous. Tergites I to III with basal carinae (62;2) diminishing in length; without basal setate depressions; tergites I to IV subequal, slightly diminishing in length. Sternite II without lateral foveae, with a single median longitudinal line and a pair of lateral lines. *Minibi* is most closely related to *Trimioarcus* and *Dalmosanus*. *Trimioarcus* is without an epipleural stria on the elytron (60;2), and basal carinae are limited to tergite I, whereas *Minibi* has the epipleural stria and basal carinae on tergites I to III. *Dalmosanus* is without basal carinae on the tergites, and is the only genus of the three to have a subhumeral fovea on the elytron.

Allomelba
Park, 1954a:16–17
(Plate 63, figs. 1–7)

Type-species: *Allomelba antennata* Park, 1954a. American Mus. Novitates no. 1674:17–19. Holotype ♂, South Bimini Island, Bahamas, British West Indies (AMNH). Monobasic.

Distribution: This monotypic genus is known only from the type locality.

Diagnosis: *Allomelba* and *Latomelba* have lost the median mesosternal fovea (63;5 and 64;4), and the procoxal foveae are minute (63;1 and 64;1). The tergites of *Allomelba* are subequal in length, but tergites I and

IV of *Latomelba* are longest. *Trimiodina* is also without a median mesosternal fovea, but differs from *Allomelba* and *Latomelba* by not having metasternal foveae.

Latomelba
Park, 1955:111
(Plate 64, figs. 1-9)

Type-species: *Latomelba quadrisicca* Park, 1955. Bull. Chicago. Acad. Sci. 10:112-113. Holotype ♂, Morce's Gap, St. Andrew Parish, Jamaica (FMNH). Monobasic.

Distribution: This genus is represented by 2 species from Jamaica.

Diagnosis: The large lateral mesosternal foveae; nonconvergent metasternal foveae (64;4); antennal segment XI lacking an apical depression; and flattened setae and truncate basal carinae of tergite I (64;8) of *Latomelba* are characters that distinguish it from *Allomelba*, as well as the differences in tergite length discussed under the latter genus (above). The presence of foveae on the metasternum of *Latomelba* (64;4) distinguishes it from *Trimiodina* (65;4).

Trimiodina
Raffray, 1898:231
(Plate 65, figs. 1-9)

Type-species: *Trimium concolor* Sharp, 1887. Biologia Centrali-Americana Coleoptera, Vol. 2, Part 1:37. Lectotype ♂, Totonicapam, 8,500-10,000 ft., Guatemala (BMNH). Monobasic.

Distribution: This monotypic genus is known only from the type locality.

Diagnosis: The absence of a median mesosternal fovea, and the triangular pattern on the mesosternum of *Trimiodina* (65;4), show a relationship to *Allomelba*, *Latomelba*, and *Trimiovillus*. The absence of foveae on the metasternum of *Trimiodina* (65;4) will distinguish it from these related genera.

The lectotype was selected from six of eleven syntypes loaned us from the British Museum of Natural History and mounted on a slide. The lectotype compares favorably with a cardboard-mounted specimen from Paris (MNHN) labeled "*Trimiodina concolor*, Type-Raffray det." which we presumed to refer to the type of the genus. No type designation for this species was found in the literature.

Quadrelba
Park, 1955:116
(Plate 66, figs. 1-7)

Type-species: *Trimiopsis parmata* Reitter, 1883. Deutsche Entomol. Zeitschr. 27:40-41. St. Thomas, Virgin Islands. Original designation.
Melba parmata (Reitter). Raffray, 1898. Rev. d'Entomol. 17:237.
Melba (Quadrelba) parmata (Reitter). Park, 1942. Northwestern Univ. Stud. Biol. Sci. Med. no. 1:120.

Distribution: The 3 species placed in this genus are found in Puerto Rico (2) and St. Thomas, Virgin Islands (3).

Diagnosis: *Quadrelba* and *Malleoceps* are without a median mesosternal fovea, as are *Allomelba* and *Latomelba*. *Quadrelba* and *Latomelba* also have widely separated metacoxae and truncate basal carinae on tergite I, but differ in the pattern of the mesosternal area (64;4 and 66;4) and lengths of tergites I and IV (Table 5). *Quadrelba* and *Malleoceps* both have anterolateral extensions of the head (66;1 and 67;2) that are quite distinctive. The metacoxae of *Quadrelba* are separated, whereas the metacoxae of *Malleoceps* are contiguous. The generic characters were based on a specimen from the Field Museum of Natural History identified as *Melba parmata* by Raffray.

Malleoceps

Park, 1954b:1

(Plate 67, figs. 1–7)

Type-species: *Malleoceps darlingtoni* Park, 1954b. Natur. Hist. Misc. no. 138:1–4. Holotype ♂, Sanchez, Dominican Republic (missing). Monobasic.

Distribution: This genus is known from the type specimen and a second species from Puerto Rico.

Diagnosis: The broad anterolateral projections that cause the head (67;2) of the male to be nearly as wide as the elytra readily distinguish this genus.

Trimiovillus

Park, 1954a:11

(Plate 68, figs. 1–7)

Type-species: *Trimiovillus bahamicus* Park, 1954a. American Mus. Novitates no. 1674:12–13. Holotype ♀, South Bimini Island, Bahamas, British West Indies (AMNH). Monobasic.

Distribution: Known only from the type specimen and a second species from Jamaica.

Diagnosis: *Trimiovillus* is considered to be relatively advanced by the absence of a median mesosternal fovea (68;5), procoxal foveae (68;1), and subhumeral fovea (68;3) of the elytron. The pattern of the mesosternal area (68;5) is very distinctive, and the presence of a median longitudinal impression on the pronotum (68;3) is unique to this genus of Part III.

Trimiomelba

Casey, 1897:563

(Plate 69, figs. 1–7)

Type-species: *Trimium convexulum* LeConte, 1878. Proc. Am. Philos. Soc. 17:383. Holotype ♀, no. 6182, Tampa, Fla. (MCZ). Designated by Bowman, 1934:144.

Distribution: This genus is limited to 3 species found on the east coast of the United States.

Diagnosis: *Trimiomelba* is the first of a cluster of 6 genera (Diagram 6) showing a marked reduction in a number of characters. All are without lateral mesocoxal foveae, procoxal foveae, and subhumeral fovea of the elytron. The more advanced have lost the median mesosternal fovea. *Trimiomelba* is closely related to *Allotrimium*, and *Melba* and is not separated from them by the characters used in Table 5. Tergite I of *Trimiomelba* (69;7) has two narrow median basal carinae and two wider lateral carinae, whereas *Allotrimium* and *Melba* are limited to a single pair of carinae on tergite I. Eight sensory depressions appear on the profemur of *Trimiovillus* (68;6); five on the profemur of *Allotrimium*; and four on the profemur of *Melba* (71;6).

Allotrimium

Park, 1943:187

(Plate 70, figs. 1–7)

Type-species: *Allotrimium michoacanensis* Park, 1943. Bull. Chicago Acad. Sci. 7:187. Holotype ♂, Zamora, Michoacan, Mexico (FMNH). Monobasic.

Distribution: Known only from the type locality.

Diagnosis: *Allotrimium*, *Trimiomelba*, and *Melba* are not separated by the characters of Table 5. The holotype of *Allotrimium michoacanensis* has widely separated, greatly reduced metasternal foveae (70;4), but these foveae were absent from another specimen. The metasternal foveae of *Trimiomelba* and *Melba* are closer and larger. The bases of the carinae of tergite I of *Allotrimium* are quite wide, but are not as widely separated as those of *Trimiomelba* (69;7).

Melba

Casey, 1897:565–566

(Plate 71, figs. 1–6; Plate 77, fig. 4; Plate 78, fig. 4; Plate 79, fig. 3)

Type-species: *Trimium thoracicum* Brendel, 1889. Entomologica Americana 5:196–197. Lectotype ♀, no. 8287, Iowa (ANSP). Designated by Bowman, 1934:144.

Distribution: This is a very large genus, with members in North, Central, and South America and the West Indies. Park has divided it into several subgenera. The nominate subgenus contains 14 species from North America, 7 from Central America and the West Indies, and 1 from Argentina. A complete revision of the genus is needed.

Diagnosis: *Melba* is closely related to *Zonaira*, *Allotrimium*, and *Trimiomelba*. *Zonaira* has basal carinae on tergites I to III, but basal carinae are limited to tergite I of the other three genera. Differences in shape and number of these carinae are discussed under *Allotrimium* and *Trimiomelba* (above). Setate depressions on the profemur of *Melba* are limited to four (71;6); but five are found on *Allotrimium*, and six to eight on *Trimiomelba*.

Zonaira, new genus

(Plate 72, figs. 1–11)

Type-species: *Zonaira trilinea*, new species.

Distribution: *Zonaira* is known only from the type species in Arizona.

Diagnosis: Vertexal and gular foveae present, with both sets of foveae opening separately. Venter of head with capitate setae; without longitudinal median carina. Antennal club of three segments (72;3). Pronotum without median discal or lateral depressions; with biarcuate antebasal sulcus (72;2); lateral foveae absent; basal depressions absent. Elytra with sutural and one discal fovea; epipleural stria present, but subhumeral fovea absent (72;2). Prosternum (72;1) without procoxal foveae, anterior prosternal foveae, or median longitudinal carina. Mesosternum (72;5) with a pair of lateral mesosternal foveae and a single median mesosternal fovea; lateral mesocoxal and promesocoxal foveae absent; small metasternal foveae present. Mesocoxal cavities open; metacoxae contiguous; profemur with sensory modification of four setae in depressions. Tergite I longest; I to III with basal carinae. Sternite II without lateral foveae; with a single median longitudinal line and a pair of lateral lines.

Zonaira has basal carinae on tergites I to III, as do *Minibi* and *Perimelba*. *Zonaira* is separated from *Minibi* by not having procoxal or lateral mesosternal foveae, and from *Perimelba* by having a median mesosternal fovea.

Zonaira trilinea, new species

(Plate 72, figs. 1–11)

Male (holotype): Head 142μ long, 260μ wide; vertexal foveae 96μ between centers. Ventral surface with simple setae, gular foveae open separately, without median carina. Antenna 316μ long, segment I 44μ long x 33μ wide; II 46μ x 33μ, III 22μ x 23μ; IV to VI 13μ x 22μ; VII 14μ x 22μ; VIII 15μ x 22μ; IX 15μ x 32μ; XI 106μ x 73μ, nearly symmetrical, with tubular and normal setae (72;3).

Pronotum 271μ long, 295μ wide. Elytron 384μ long, 218μ wide, sutural and antebasal foveae with several guard setae. Profemur 277μ long x 70μ wide, sensory area with four depressions; mesofemur (72;7) with mesial projection; mesofemur 271μ long x 73μ wide; mesotibia (72;8) with curved spine 90μ from apex; metacoxa (72;10) with short blunt spine; metafemur 295μ long x 58μ wide.

First visible tergite 330μ at base; basal carinae 29μ long, 129μ between bases; tergite II with basal carinae 15μ long, 102μ between bases; tergite III with carinae 15μ long, 94μ between base. Sternite II with lateral basal carinae 22μ long, 265μ between bases, with narrow median carina 15μ long; sternite III with lateral setate projection (72;6). Penial plate nearly symmetrical, 44μ long x 80μ wide. Genitalia (72;11) 171μ long x 65μ deep.

Female: Abdominal segments VIII and IX as illustrated in Plate 72; figure 9.

Holotype and allotype are slide-mounted. Point-mounted paratypes are red-brown. One ♀ measures 1.25 mm in length, .45 mm in width.

Distribution: The holotype male and four paratype females were collected in the Santa Catalina foothills, Arizona, on April 6, 1968, by K. Stephan. The type (no. 905) and paratypes are deposited at Davis (UCD).

Perimelba, new status
Park, 1943:183

(Plate 73, figs. 1–6)

Type-species: *Melba* (*Perimelba*) *granulosa* Park, 1943. Bull. Chicago Acad. Sci. 7:184–185. Holotype ♀, Victoria, Tamaulipas, Mexico (FMNH). Designated by Park, 1952. Chicago Acad. Sci. Spec. Pub. no. 9, Part 2:141.

Distribution: This genus is reported to contain 8 species; 5 are found in Central America, 1 in Jamaica, 1 in Argentina, and 1 in California and Arizona.

Diagnosis: The absence of a median mesosternal fovea (73;4) separates *Perimelba* from *Zonaira*. *Hanfordia* also is without a median mesosternal fovea, but *Perimelba* differs from *Hanfordia* by having a pair of metasternal foveae (73;4).

Hanfordia
Park, 1960:10

(Plate 74, figs. 1–8)

Type-species: *Hanfordia absoluta* Park, 1960. Trans. Am. Microscop. Soc. 79:10–11. Holotype ♂, Cooper's Hill, Red Hills, St. Andrew Parish, Surrey County, Jamaica (FMNH). Monobasic.

Distribution: This is a monotypic genus thus far found only at the type locality.

Diagnosis: *Hanfordia* is the most advanced genus in this cluster. It is without foveae or sulci on the prothorax (74;1,3) and limited to a sutural stria on the elytron (74;3). The mesosternum has but a pair of lateral mesosternal foveae (74;5).

GENERA OF QUESTIONABLE POSITION ASSOCIATED WITH GROUP C

Actionoma
Raffray, 1898:235

Type-species: *Actionoma obesum* Raffray, 1898. Rev. d'Entomol. 17:235. Mexico. Monobasic.

Distribution: This monotypic genus is reported only from the type locality.

We have not seen representatives of this genus, and it is placed with this group only because Raffray stated that it resembled *Actium*.

Basolum
Casey, 1897:571

Type-species: *Trimium impunctatum* Brendel 1890. *In* Brendel and Wickham. Bull. Lab. Natur. Hist., Univ. Iowa 2:34. Virginia. Designated by Bowman, 1934:144.

Distribution: Brendel and Wickham report *impunctatum* to be from Virginia and Maryland.

Casey established *Basolum*, with some question, for 2 species, based on their descriptions, as no specimens were available to him. We have been unable to locate the type specimens.

Dalmoplectus
Raffray, 1890:96, 102

Type-species: *Dalmodes rybaxides* Reitter, 1882b. Verh. k. k. zool.-bot. Gesell. Wein 32:382. Mexico. Designated by Park, 1952:128.

Distribution: The type species from Mexico and a second species from Brazil comprise this genus, but there is some question concerning the generic status of the second species (see Park 1952:128).

We have not seen *D. rybaxides*, and it is placed with those species of Group C (probably Part III) because Park (1952) associated it with genera we have assigned to this group.

Pseudotrimium
Raffray, 1898:230

Type-species: *Pseudotrimium microcephalum* Raffray, 1898. Rev. d'Entomol. 17:230 (MNHN). Monobasic.

Distribution: The type locality and distribution of this monotypic genus is somewhat in doubt, since the author lists it from Yucatan, Téapa Mexico, and "New Orleans."

We have seen this species and tentatively place the genus in Group C. Park (1953) associated it with *Dalmosella*. We were not able to slide-mount the specimen to determine the foveation for a more positive group assignment.

Ramelbida
Park, 1942:112

Type-species: *Melba quadrifoveata* Raffray, 1903. Ann. Soc. Entomol. France 72:537. St. Thomas, Virgin Islands. Monobasic.

Distribution: This monotypic genus is known only from the type locality.

We have not seen this species. Park reports it to be similar to *Trimiosella*, but differing by having four vertexal foveae. This is indeed different!

Trimiosella
Raffray, 1898:236

Type-species: *Trimiopsis anguina* Reitter, 1883. Deutsche Entomol. Zeitschr. 27:42. St. Thomas, Virgin Islands (MNHN). Monobasic.

Distribution: This is a monotypic genus known only from the type locality.

A point-mounted specimen from the Paris Museum is similar to those genera placed in Part III. No vertexal foveae were evident, but a slide-mount would be necessary to confirm this and other characters in order to make a definite assignment to Group C.

Zolium
Casey, 1897:560

Type-species: *Trimiopsis eggersi* Reitter, 1883. Deutsche Entomol. Zeitschr. 27:38. St. Thomas, Virgin Islands. Monobasic.
Melba eggersi (Reitter). Raffray, 1898. Rev. d'Entomol. 17:237.
Zolium eggersi (Reitter). Park, Wagner, and Sanderson, 1976. Fieldiana Zool. 68:45.

Distribution: This monotypic genus is known only from the type locality.

This genus was considered a synonym of *Melba* by Park in 1943, but it was given generic status by Park *et al.* in 1976 without reasons for the change. We have not seen the genus.

ADDENDUM

Following the completion of our plates and diagrams, 3 new genera of Euplectini were described from the West Indies.

Hispanisella
Park, 1976:32

Type-species: *Hispanisella haitiana* Park, 1976. *In* Park, Wagner, and Sanderson. Fieldiana Zool. 68:33. Holotype ♂, Refuge, Haiti (FMNH). Original designation.

Distribution: The genus is composed of 2 species from Haiti.

Diagnosis: The absence of both a median mesosternal fovea and foveae on sternite II on *Hispanisella* are characters shared with 7 genera of Group C, Part III. In Table 5, *Hispanisella* would be placed near *Quadrelba*, as it shows the same characters listed for *Quadrelba*, except that it is without basal carinae on tergite I. This character can also be used to separate *Hispanisella* in our bracket key. In Diagram 6, *Hispanisella* would be placed near *Latomelba* and in the same cluster with *Malleoceps* and *Quadrelba*. It differs from *Latomelba* by not having tergite IV as long as I, and from all three of these genera by the absence of carinae on the tergites. Basal carinae are also absent on *Haasiella*, but this genus has a subhumeral fovea.

Haasiella
Park, 1976:35–36

Type-species: *Haasiella medicina* Park, 1976. *In* Park, Wagner, and Sanderson. Fieldiana Zool. 68:36,38. Holotype ♂, Luquillo Forest, 1,750 ft., Puerto Rico (FMNH). Monobasic.

Distribution: This genus is known only from the type specimen.

Diagnosis: *Haasiella* is also placed in Group C, Part III, for the same reasons discussed for *Hispanisella* (above). It is compared with *Lemelba* by Park et al., but is more advanced in our proposed phylogeny by not having a median mesosternal fovea. *Haasiella* shows the same combination of characteristics in Table 5 as *Quadrelba*, except that *Haasiella* does not have carinae on tergite I, and the elytron of *Haasiella* has a subhumeral fovea. The procoxal foveae of *Haasiella* are very small. *Haasiella* goes to Group C, couplet 10, in the bracket key, but differs from *Bibloplectus* by the elytron having one antebasal fovea, and from the alternative by having a subhumeral fovea. The placement of *Haasiella* in diagram 6 would be between *Latomelba* and *Quadrelba*.

Sandersonella
Park, 1976:38

Type-species: *Sandersonella transversa* Park, 1976. *In* Park, Wagner, and Sanderson. Fieldiana Zool. 68:39. Holotype ♂, Tombeau Cheval, Haiti (INHS). Monobasic.

Distribution: This genus is represented only by the type specimen.

We have not been able to see the type. It is tentatively placed in Group C because Park et al. (1976) allied it with *Latomelba*, *Quadrelba*, and *Allomelba*.

References Cited

BOWMAN, J. R.
   1934.   The Pselaphidae of North America. Published privately, Pittsburgh, Pa. vi + 149pp.
BRENDEL, E.
   1889.   Descriptions of new Scydmaenidae and Pselaphidae. Entomologica Americana 5:193–197.
   1892.   *Rhexidius*. Entomol. News 3:11–13.
BRENDEL, E., and H. F. WICKHAM
   1890.   The Pselaphidae of North America. Bull. Lab. Natur. Hist., Univ. Iowa 1:216–304; 2:1–84.
CASEY, T. L.
   1884.   Contributions to the Descriptive and Systematic Coleopterology of North America, Parts I and II. Published privately, Collins House, Philadelphia. 198 pp.
   1886.   Descriptive notices of North American Coleoptera, I. Bull. Calif. Acad. Sci. 2:157–264.
   1887.   On some new North American Pselaphidae. Bull. Calif. Acad. Sci. 2:455–482.
   1893.   Coleopterological Notices, V. Ann. New York Acad. Sci. 7:281–606.
   1897.   Coleopterological Notices, VII. Ann. New York Acad. Sci. 9:285–684.
   1908.   Remarks on some new Pselaphidae. Canadian Entomol. 40:257–281.
CHAUDOIR, M.
   1845.   I. Pselaphides. Bull. Soc. Imper. Naturalistes de Moscou 18:164–181.
DENNY, H.
   1825.   Monographia Pselaphidarum ed Scydmaenidarum Britanniae. Norwich, Great Britain. vi + 72 pp. + 14 pl.
FLETCHER, F. C.
   1928.   Pselaphidae collected by Dr. Alfons Dampf in Central America. Ann. Entomol. Soc. America 21:203–231.
GRIGARICK, A. A., and R. O. SCHUSTER
   1966.   A new genus of the tribe Euplectini in California. Pan-Pacific Entomol. 42:31–33.
   1968.   A revision of the genus *Cupila* Casey. Pan-Pacific Entomol. 44:38–44.
   1970.   A new genus in the tribe Euplectini. Pan-Pacific Entomol. 46:36–39.
   1976.   A revision of the genus *Oropodes* Casey. Pan-Pacific Entomol. 52:97–109.
LEACH, W. E.
   1817.   On the stirpes and genera composing the family Pselaphidae; with the names of British species. The Zool. Misc. 3:80–87.
LeCONTE, J. L.
   1850.   On the Pselaphidae of the United States. Boston Jour. Natur. Hist. 6:64–110.
   1863.   New species of North American Coleoptera, Part I. Smithsonian Misc. Collections VI, no. 167:1–92.

    1878.   Additional descriptions of new species. Proc. Am. Philos. Soc. 17:373–434.

    1880.   Short studies of North American Coleoptera. Trans. Am. Entomol. Soc. 8:163–218.

MÄKLIN, F. G.

    1852.   *In* Mannerheim, Zweiter Nachtrag Zur Kaefer-Fauna der Nord-Americanischen Laender des Russischen Reiches. Bull. Soc. Imper. Naturalistes de Moscou 25:283–372.

PARK, O.

    1942.   A study of Neotropical Pselaphidae. Northwestern Univ. Stud. Biol. Sci. Med. no. 1. ix + 403 pp.

    1943.   A preliminary study of the Pselaphidae (Coleoptera) of Mexico. Bull. Chicago Acad. Sci. 7:171–226.

    1945.   Further studies in Pselaphidae of Mexico and Guatemala. Bull. Chicago Acad. Sci. 7:331–443.

    1949.   New species of Nearctic pselaphid beetles and a revision of the genus *Cedius*. Bull Chicago Acad. Sci. 8:315–343.

    1952.   A revisional study of Neotropical pselaphid beetles. Chicago Acad. Sci. Spec. Pub. no. 9, Part 2:53–150.

    1953a. A European pselaphid beetle collected in New York. Natur. Hist. Misc. no. 117:1–3.

    1953b. A new genus of pselaphid beetles from the Everglades. Natur. Hist. Misc. no. 121:1–4.

    1954a. The Pselaphidae of South Bimini Island, Bahamas, British West Indies. American Mus. Novitates no. 1674:1–25.

    1954b. A new genus of pselaphid beetles from the Antilles. Natur. Hist. Misc. no. 138:1–4.

    1955.   Contribution to the pselaphid beetle fauna of Jamaica. Bull. Chicago Acad. Sci. 10:101–122.

    1960.   Pselaphid beetles of Jamaica. Trans. Am. Microscop. Soc. 79:5–25.

    1963.   Observations on the genus *Actium*. Trans. Am. Microscop. Soc. 82:178–184.

PARK, O., and R. O. SCHUSTER

    1955.   A new subtribe of pselaphid beetles from California. Natur. Hist. Misc. no. 148:1–6.

PARK, O., and J. A. WAGNER

    1961.   The family Pselaphidae. *In* M. H. Hatch, The beetles of the Pacific Northwest, Part III: Pselaphidae and Diversicornia I. Univ. Wash. Pub. Biol. 16:1–380.

PARK, O., J. A. WAGNER, and M. W. SANDERSON

    1976.   Review of the pselaphid beetles of the West Indies. Fieldiana Zool. 68:XI + 1–90.

RAFFRAY, A.

    1890.   Étude sur les Psélaphides: Genera et descriptions d'espèces nouvelles. Rev. d'Entomol. 9:1–264.

    1898.   Notes sur les Psélaphides. Rev. d'Entomol. 17:198–273.

    1903– Genera et catalogue des Psélaphides. Ann. Soc. Entomol. France 72:484–604 (1903);

    1904.   73:1–476, 636–658 (1904).

    1908.   Supplement a la liste des Coleopteres de la Guadeloupe (Pselaphidae), 2nd Suppl. Ann. Soc. Entomol. France 77:33–40.

    1911.   Pselaphidae. Junk's Coleopterorum Catalogus, Part 27:1–222.

REICHENBACH, H. T. L.

    1816.   Monographia Pselaphorum. Lipsiae. 80 pp. + 2 pl.

REITTER, E.

    1881.   Bestimmungus-Tabellen der europäischen Coleopteren Verh. k. k. zool.-bot. Gesell. Wein 31:443–592.

    1882a. Neue Pselaphiden und Scydmaeniden aus Brasilien. Deutsche Entomol. Zeitschr. 26:129–152.

1882b. Neue Pselaphiden und Scydmaeniden aus Central und Sudamerika. Verh. k. k. zool.-bot. Gesell. Wein 32:371–386.

1883. Beitrag zur Kenntniss der Clavigeriden, Pselaphiden und Scydmaeniden von Westindien. Deutsche Entomol. Zeitschr. 27:33–46.

SCHUSTER, R. O., and A. A. GRIGARICK

1968. A new genus of pselaphid beetle from southeastern United States. Pan-Pacific Entomol. 44:112–118.

SCHUSTER, R. O., and G. A. MARSH

1957. A new genus of Euplectini from California. Pan-Pacific Entomol. 33:149–152.

SHARP, D.

1887. Pselaphidae. Biologia Centrali-Americana Coleoptera, Vol. 2, Part 1:1–46.

THOMSON, C. G.

1861. Skandinaviens Coleoptera. Pselaphidae 3:220–240.

WAGNER, J. A.

1975. Review of the genera *Euplectus*, *Pycnoplectus*, *Leptoplectus*, and *Acolonia* including Nearctic species north of Mexico. Entomologica Americana 49:125–207.

Index to Genera

Bracket Key to Groups and Genera

The characters used in the Key to Genera were selected for their simplicity and brevity to facilitate rapid identification. For the same reasons single characters are frequently used; and we suggest that until the user becomes familiar with the key, identifications be confirmed by comparisons with illustrations and the tabular keys.

Specimens simply cleared and slide-mounted, venter up, should be sufficient to provide the information necessary to use the key. Separation of the tergites from the sternites is helpful, and separate mounting of the male or female genital structures (preferably on the same slide) is requisite in order to orient these minute structures for comparative observation.

Generic names in the key are followed by the text page on which the genus is considered and by the number of the plate illustrating characters of the genus.

1	Lateral mesosternal foveae forked	Group B,	p. 24
	—Lateral mesosternal foveae not forked	2	
2(1)	Median mesosternal foveae paired; forked therminally if single; if mesosternal foveae absent, antennal foveae present	Group A,	p. 13
	—Median mesosternal foveae single and not forked, or absent	Group C,	p. 30

Group A

1	Metasternum without foveae	2	
	—Metasternum with one fovea	*Aboeurhexius*	p. 23, pl. 25
	—Metasternum with two foveae	10	
	—Metasternum with three foveae	*Rhinoscepsis*	p. 17, pl. 2
2(1)	Single forked median mesosternal fovea present	*Rhexiola*	p. 22, pl. 22
	Two median mesosternal foveae present	3	
3(2)	Lateral prosternal fovea present	*Euplectus*	p. 22, pl. 20
	—Lateral prosternal fovea absent	4	
4(3)	Elytron with subhumeral fovea	5	
	—Elytron without subhumeral fovea	9	
5(4)	Elytron with two antebasal foveae	*Rhexinia*	p. 22, pl. 21
	—Elytron with three antebasal foveae	6	
6(5)	Sternite II with median foveae	7	
	—Sternite II without median foveae	*Acolonia*	p. 19, pl. 10
7(6)	Profemur with distinct row of pores, tubercles, spines or carina	8	
	—Profemur without distinct row of pores, tubercles, spines or carina	*Pycnoplectus*	p. 18, pl. 9
8(7)	Mesocoxal cavities closed	*Euboarhexius*	p. 21, pl. 17
	—Mesocoxal cavities open	*Rhexius*	p. 21, pl. 18

9(4)   Elytron with one antebasal fovea.....................................*Trimioplectus* p. 18, pl. 8
        —Elytron with two antebasal foveae................................*Trigonoplectus* p. 18, pl. 7
10(1)   Single actual or apparent median mesosternal fovea 11
        —Two distinct median mesosternal foveae 12
11(10) Median mesosternal fovea forked ....*Hatchia* p. 23, pl. 23
        —Median mesosternal fovea not forked ....*Trisignis* p. 17, pl. 4
12(10) Sternite II with median foveae 13
        —Sternite II without median foveae....................................... 16
13(12) Basolateral margin of pronotum with single distinct spine ....*Oropus* p. 20, pl. 13
        —Basolateral margin of pronotum without spine............................. 14
14(13) With foveate depression basolateral to antennal insertion....................... 15
        —Without foveate depression basolateral to antennal insertion............*Fletcherexius* p. 20, pl. 15
15(14) Pronotum with seta arising from each tubercle; tergite I with basal carinae .. *Hexirhexius* p. 20, pl. 16
        —Pronotum without setate tubercles; tergite I without basal carinae ....*Rhexidius* p. 20, pl. 14
16(12) Tarsi of two segments... 17
        —Tarsi of three segments 18
17(16) Dorsal antennal foveae distinctly separated ....*Barroeuplectoides* p. 19, pl. 11
        —Dorsal antennal foveae with common opening ....*Tuberoplectus* p. 19, pl. 12
18(16) Prosternum with median carina 19
        —Prosternum without median carina 21
19(18) Elytron with two antebasal foveae diverging from common depression.........*Eutyphlus* p. 17, pl. 3
        —Elytron with two or three distinctly separate antebasal foveae................... 20
20(19) Profemur with distinct row of pores, tubercles or spines ....*Bibloporus* p. 23, pl. 24
        —Profemur without row of pores, tubercles or spines ....*Thesium* p. 13, pl. 1
21(18) Elytron with subhumeral fovea 22
        —Elytron without subhumeral fovea ....*Eurhexius* p. 22, pl. 19
22(21) Tergite I with basal depression and basal carinae.....................*Pseudactium* p. 17, pl. 5
        —Tergite I without basal depression or basal carinae ....*Ramecia* p. 18, pl. 6

Group B

        Metasternum with two fovea 2
        —Metasternum without foveae... 9
2(1)   Sternite II with median foveae 3
        —Sternite II without median foveae....................................... 6
3(2)   Elytron with one antebasal fovea 4
        —Elytron with two or three antebasal foveae................................. 5
4(3)   One forked median mesosternal foveae present...........................*Trichonyx* p. 27, pl. 30
        —Two unforked median mesosternal foveae present ....*Bontomtes* p. 24, pl. 27
5(3)   Ventral surface of head with large median anterior tubercle...................*Morius* p. 24, pl. 26
        —Ventral surface of head without median anterior tubercle ....*Foveoscapha* p. 27, pl. 28
6(2)   Elytron without antebasal foveae ....*Panaramecia* p. 29, pl. 37
        —Elytron with one antebasal fovea..................................*Abdiunguis* p. 28, pl. 31
        —Elytron with three antebasal foveae 7
7(6)   Pronotum with median longitudinal impression ....*Verabarolus* p. 29, pl. 35
        —Pronotum without median longitudinal impression 8
8(7)   Profemur with sensory modification; tergite lengths subequal ....*Thesiectus* p. 29, pl. 36
        —Profemur without sensory modification; tergite I longest ....*Mexiplectus* p. 28, pl. 34
9(1)   Subhumeral fovea present... 10
        —Subhumeral foveae absent ....*Tetrascapha* p. 28, pl. 33
10(9)   Promesocoxal foveae present ....*Euplecturga* p. 28, pl. 32
        —Promesocoxal foveae absent ....*Oropodes* p. 27, pl. 29

Group C

1   Median mesosternal fovea absent 2
    —Median mesosternal fovea present 14
2(1) Lateral mesocoxal foveae absent 3
    —Lateral mesocoxal foveae present 4
3(2) Metasternal foveae present . . . . *Perimelba* p. 44, pl. 73
    —Metasternal foveae absent . . . . *Hanfordia* p. 44, pl. 74
4(2) Basal carina of tergite III absent; basal depressions may be present 5
    —Basal carina of tergite III present; without basal depression . . . . *Saxet* p. 37, pl. 50
5(4) Head with ventral median carina 6
    —Head without ventral median carina 7
6(5) Procoxal fovea present . . . . *Euplectamecia* p. 36, pl. 47
    —Procoxal fovea absent . . . . *Dalmomelba* p. 37, pl. 51
7(5) Pronotum with median longitudinal impression 8
    —Pronotum without median longitudinal impression 10
8(7) Tergites I to III without basal depressions 9
    —Tergites I to III with basal depressions . . . . *Thesiastes* p. 36, pl. 48
9(8) Mesosternal area reticulate, not setate; profemur with sensory modification . *Trimiovillus* p. 42, pl. 68
    —Mesosternal area not reticulate, with numerous setae; profemur without sensory modification . . . . *Trimiopsis* p. 38, pl. 52
10(7) Elytron with one antebasal fovea, without subhumeral fovea 11
    —Elytron with two antebasal foveae; with subhumeral foveae . . . . *Bibloplectus* p. 37, pl. 49
11(10) Metasternal foveae absent . . . . *Trimiodina* p. 41, pl. 65
    —Metasternal foveae present 12
12(11) Metasternal foveae convergent, close together; antennal segment XI with two wide subapical setae set in depression; tergites subequal in length . . . . *Allomelba* p. 40, pl. 63
    —Metasternal foveae not convergent, distinctly spaced; antennal segment XI without subapical setae in depression; tergite I, or I and IV longest 13
13(12) Head uniformly convergent anterior to eyes; ventral surface of head with 14 pairs capitate setae; tergites I and IV longest . . . . *Latomelba* p. 41, pl. 64
    —Head of equal width anterior and posterior of eyes; ventral setae simple; tergite I longest . . . . *Quadrelba* p. 41, pl. 66
    —Head greatly expanded anteriorly; eyes not visible from above; ventral setae simple; tergite I longest . . . . *Malleoceps* p. 42, pl. 67
14(1) Lateral mesocoxal foveae present 15
    —Lateral mesocoxal foveae absent 30
15(14) Metasternal foveae absent 16
    —Metasternal fovea single . . . . *Allobrox* p. 38, pl. 53
    —Metasternal foveae double (pores may be present but not on suture) 19
16(15) Basal carinae on tergite I only 17
    —Basal carinae on tergites I to III . . . . *Liniolis* p. 34, pl. 41
17(16) Subhumeral fovea and epipleural stria present 18
    —Subhumeral fovea and epipleural stria absent . . . . *Dalmosella* p. 39, pl. 57
18(17) Procoxal foveae present . . . . *Actiastes* p. 38, pl. 55
    —Procoxal foveae absent . . . . *Simplona* p. 38, pl. 56
19(15) Sternite II with median foveae; pronotum with median longitudinal impression 20
    —Sternite II without median foveae; pronotum without median longitudinal impression 21
20(19) Venter of head and prosternum medianly carinate . . . . *Mexigaster* p. 30, pl. 38
    —Venter of head and prosternum not medianly carinate . . . . *Leptoplectus* p. 34, pl. 39
21(19) Metacoxae contiguous 23
    —Metacoxae widely separated 22
22(21) Mesosternal area entirely reticulate, without setae; subhumeral fovea absent . . . . *Biblomelba* p. 39, pl. 58

PLATES

PLATE 1. *Thesium cavifrons* (LeConte). Group A. Figs. 1–5, 7 ♂; fig. 6 ♀.

Fig. 1. Prosternum.
Fig. 2. Mesosternal area.
Fig. 3. Dorsal aspect.
Fig. 4. Protarsal segments II and III and claw.
Fig. 5. Tergite I.
Fig. 6. Abdominal segments VIII and IX.
Fig. 7. Genitalia, lateral aspect.

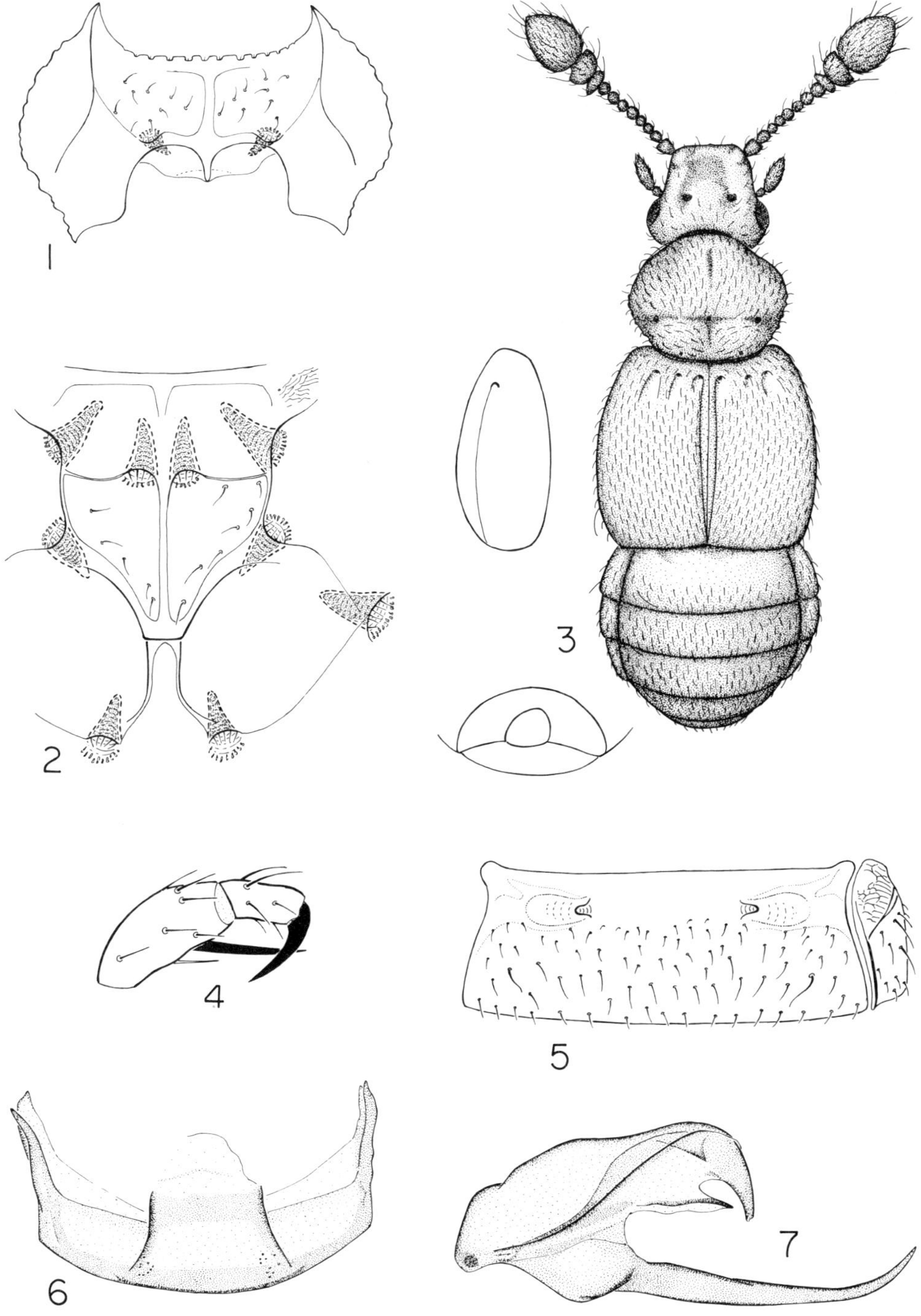

PLATE 2. *Rhinoscepsis bistriatus* LeConte. Group A. Figs. 1-3, 5-7 ♂; figs. 4, 8 ♀.

Fig. 1. Mesosternal area.
Fig. 2. Dorsal aspect.
Fig. 3. Antennal segments VIII to XI.
Fig. 4. Abdominal segments VIII and IX.
Fig. 5. Maxillary palp.
Fig. 6. Mesotarsal claws.
Fig. 7. Genitalia, lateral aspect.
Fig. 8. Profemur.

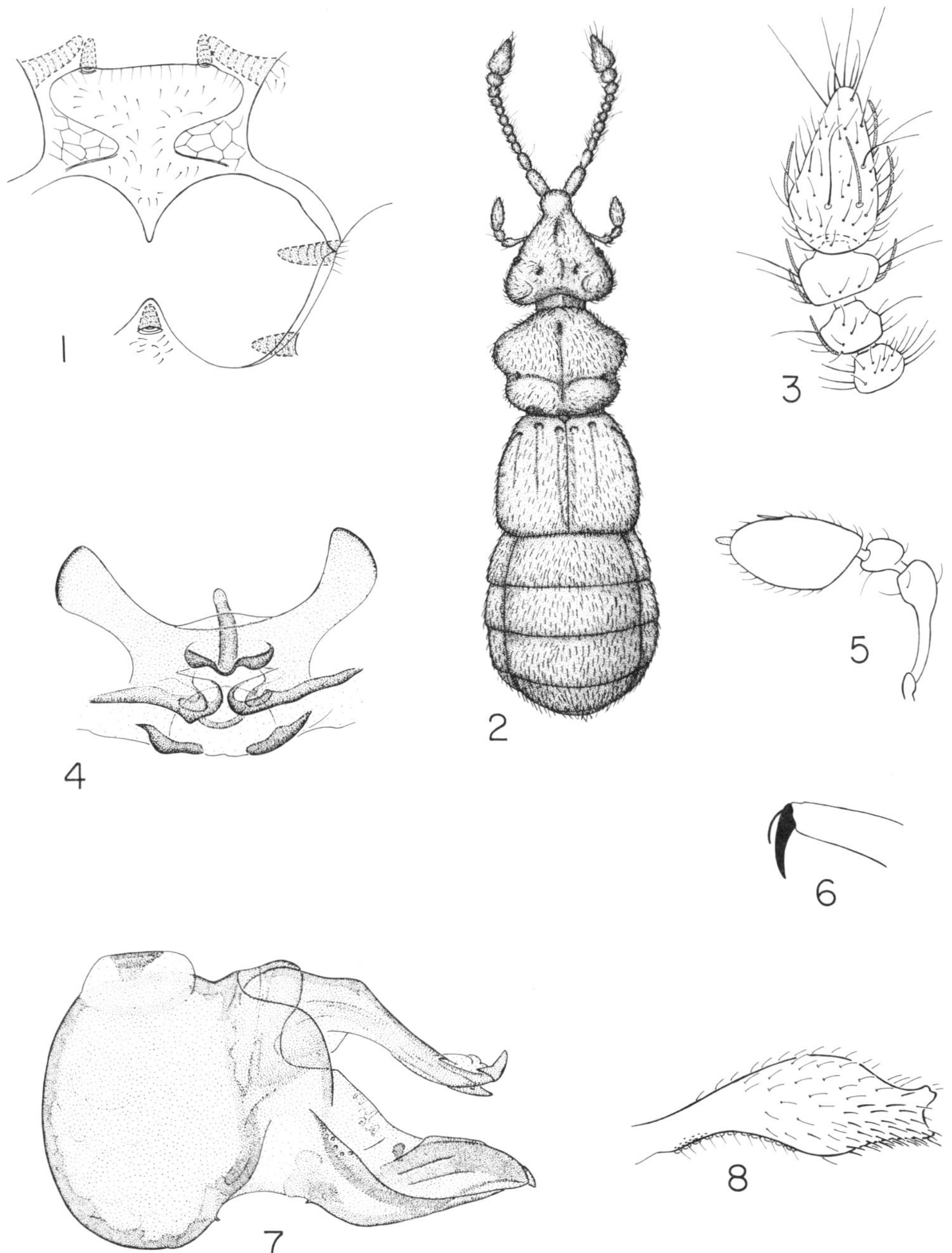

PLATE 3. *Eutyphlus similis* LeConte. Group A. Homotype. Figs. 1-6, 8 ♂; fig. 7 ♀.

Fig. 1. Metatarsal claw.
Fig. 2. Prosternum.
Fig. 3. Dorsal aspect.
Fig. 4. Antennal segments IX to XI.
Fig. 5. Mesosternal area.
Fig. 6. Tergite I.
Fig. 7. Abdominal segments VIII and IX.
Fig. 8. Genitalia, lateral aspect.

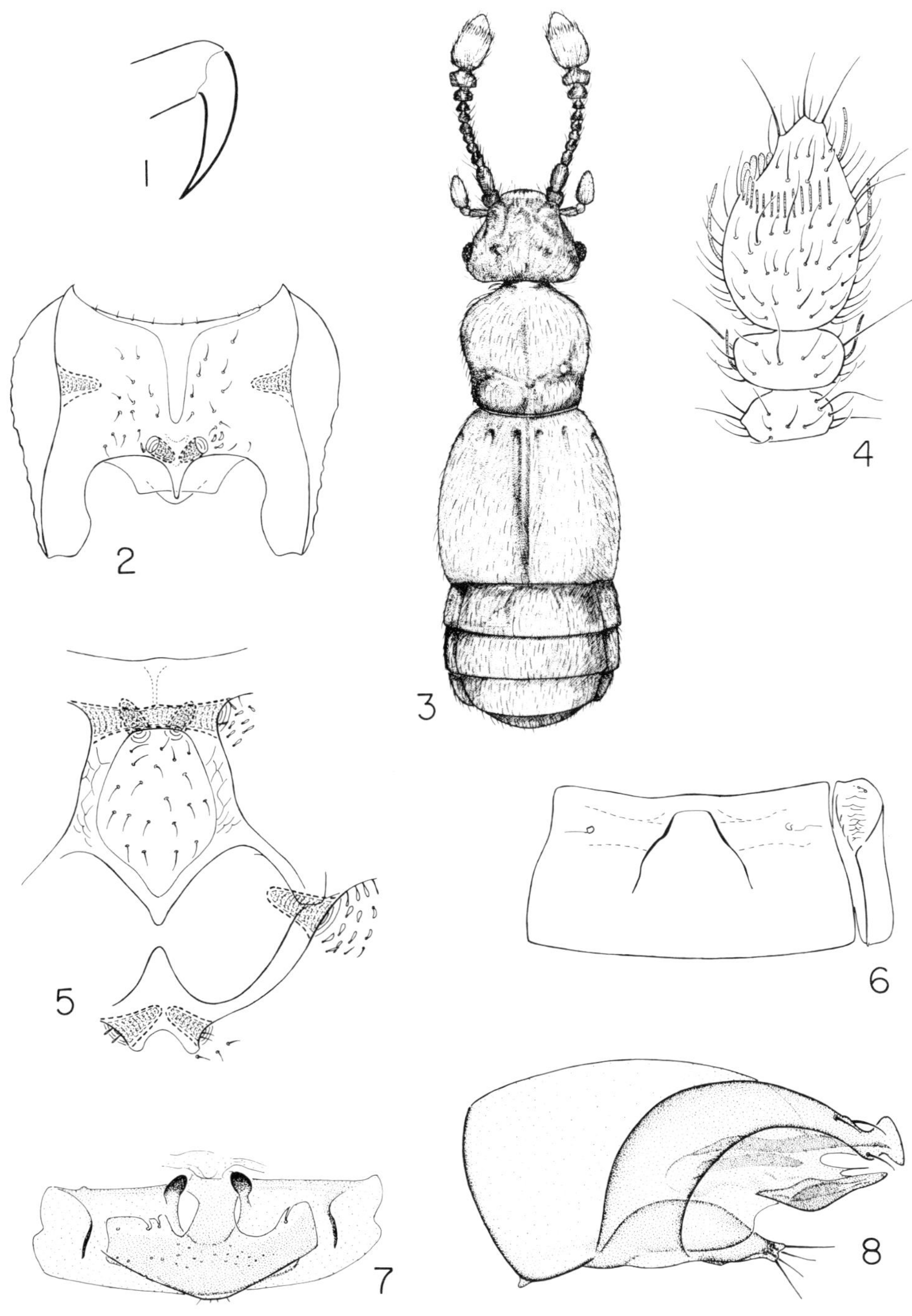
1
2
3
4
5
6
7
8

PLATE 4. *Trisignis helferi* Park and Schuster. Group A. Figs. 1, 2, 4-6, 8 ♀; figs. 7, 9 ♂. *Trisignis marshi* Park and Schuster. Fig. 3 ♂.

Fig. 1. Arrangement of gular fovea, tentoria, and ventral foveae.
Fig. 2. Prosternum.
Fig. 3. Dorsal aspect.
Fig. 4. Antennal segments IX to XI.
Fig. 5. Mesosternal area.
Fig. 6. Metatarsal claws.
Fig. 7. Penial plate.
Fig. 8. Abdominal segments VIII and IX.
Fig. 9. Genitalia, lateral aspect.

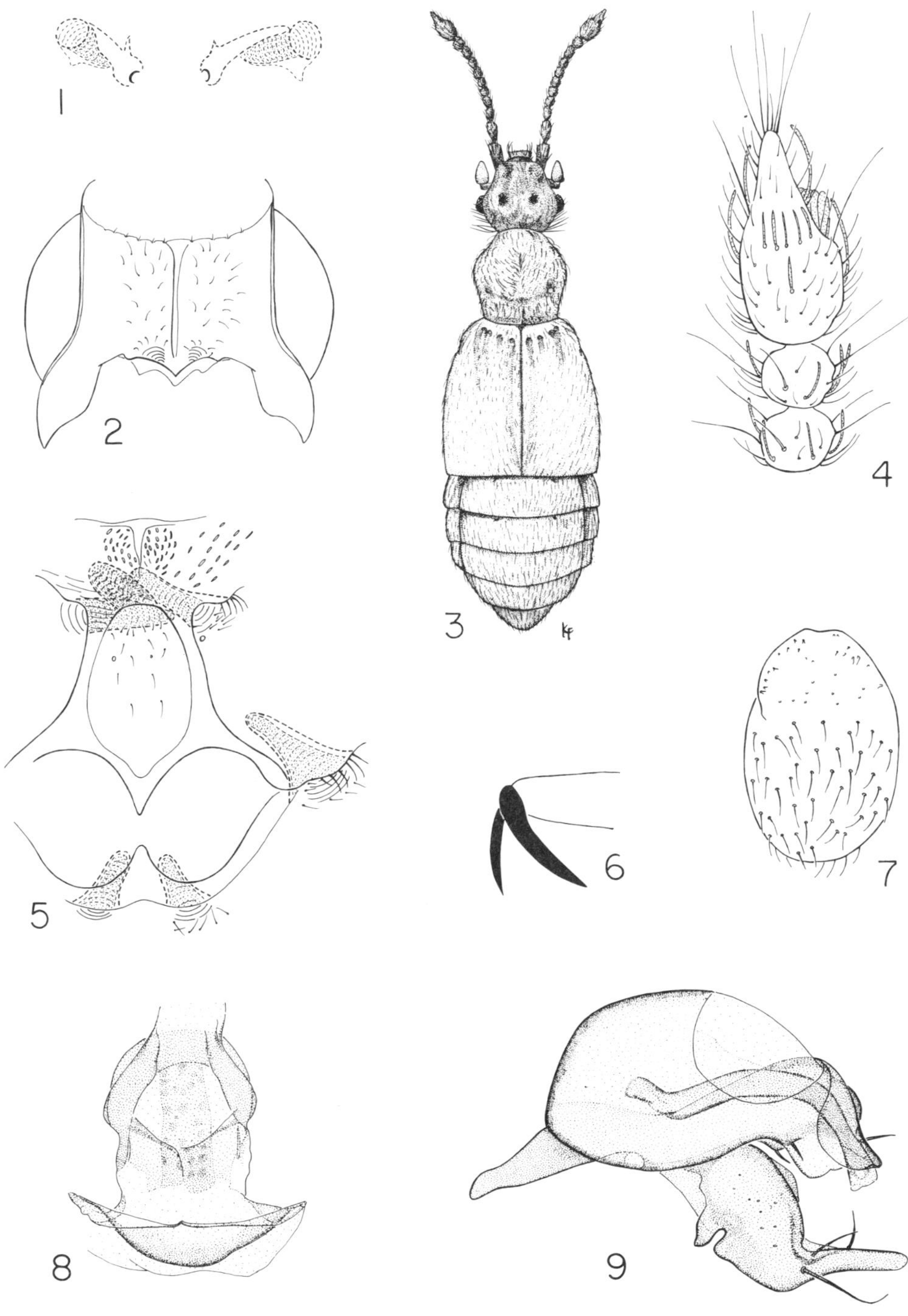

PLATE 5. *Pseudactium carolinae* Casey. Group A. Type ♀. Fig. 2. *Pseudactium cephalicum* Casey. Type ♀. Figs. 1, 3-7.

Fig. 1. Mesotarsal claws.
Fig. 2. Dorsal aspect.
Fig. 3. Antennal segments VII to XI.
Fig. 4. Sternite VI.
Fig. 5. Prosternum.
Fig. 6. Mesosternal area.
Fig. 7. Abdominal segments VIII and IX.

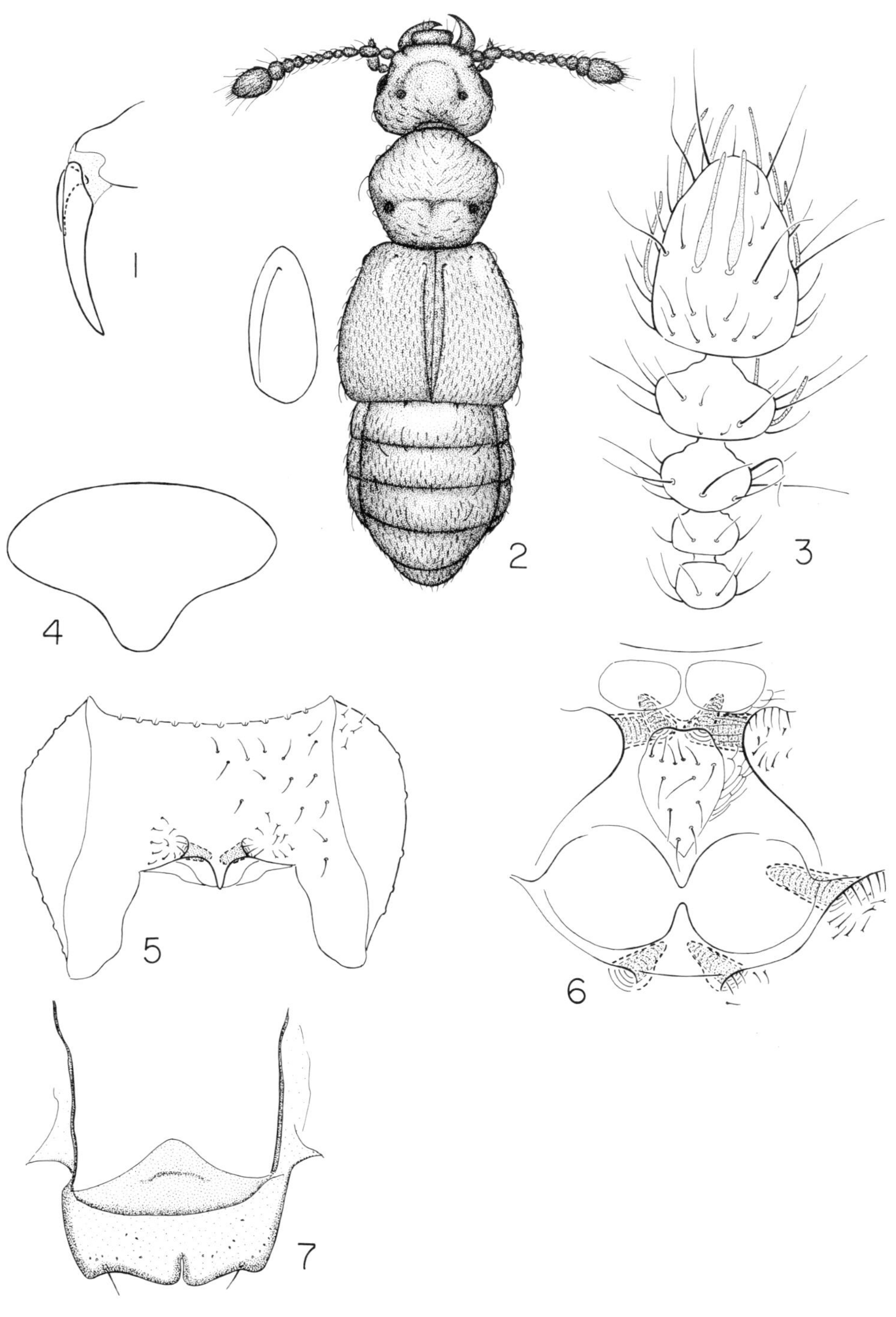

PLATE 6. *Ramecia arcuata* (LeConte). Group A. ♂.

Fig. 1. Prosternum.
Fig. 2. Mesofemur—trochanter articulation.
Fig. 3. Dorsal aspect.
Fig. 4. Antennal segments VII to XI.
Fig. 5. Mesosternal area.
Fig. 6. Mesotarsal claw.
Fig. 7. Sternites I to III, lateral margins.
Fig. 8. Profemur.
Fig. 9. Genitalia, dorsal aspect.

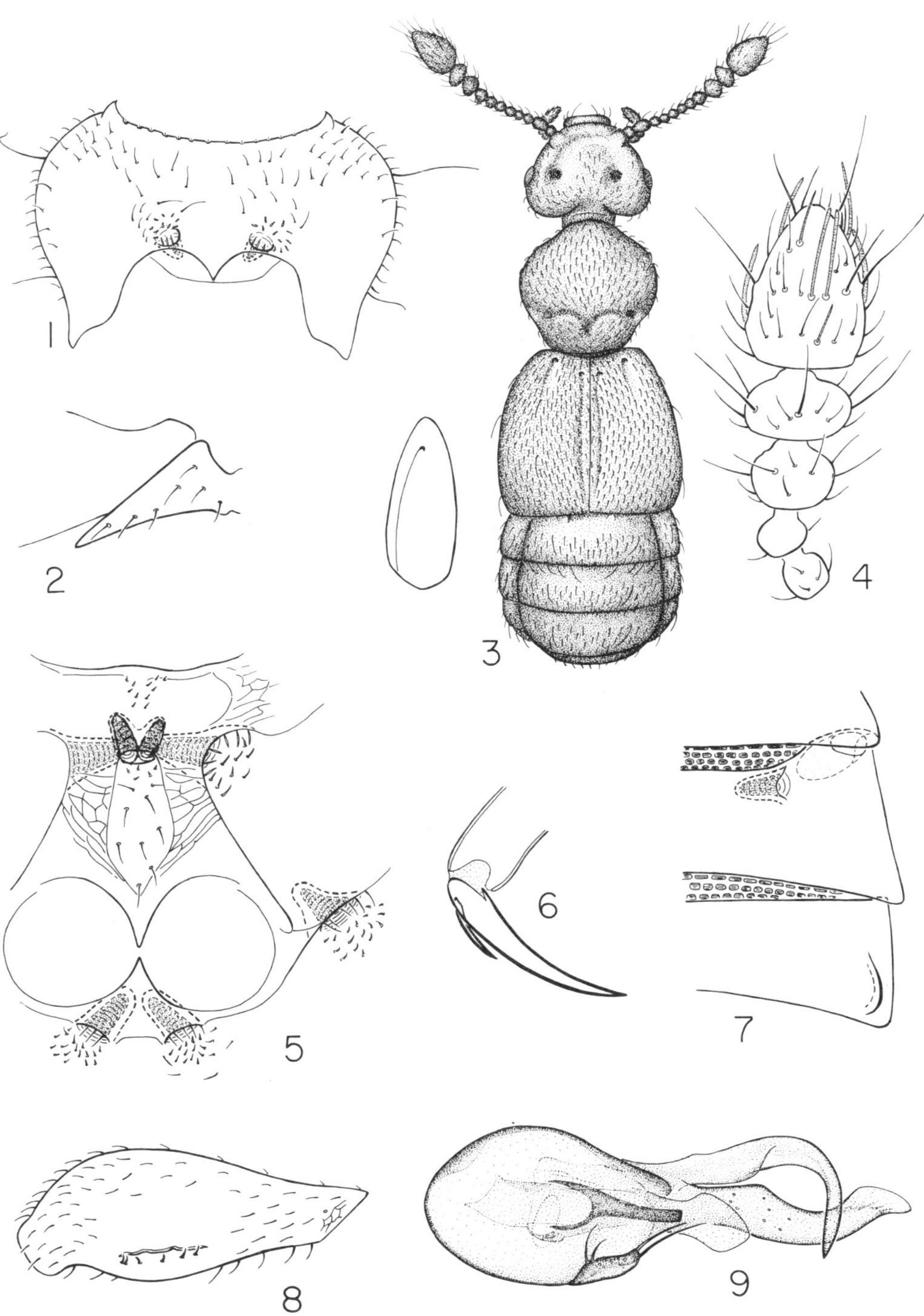

PLATE 7. *Trigonoplectus minutus* Bowman. Group A. Type ♂.

Fig. 1. Prosternal area.
Fig. 2. Dorsal aspect.
Fig. 3. Antennal segments VIII to XI.
Fig. 4. Mesosternal area.
Fig. 5. Mesotarsal claw.
Fig. 6. Tergite I.
Fig. 7. Profemur.
Fig. 8. Genitalia, lateral aspect.

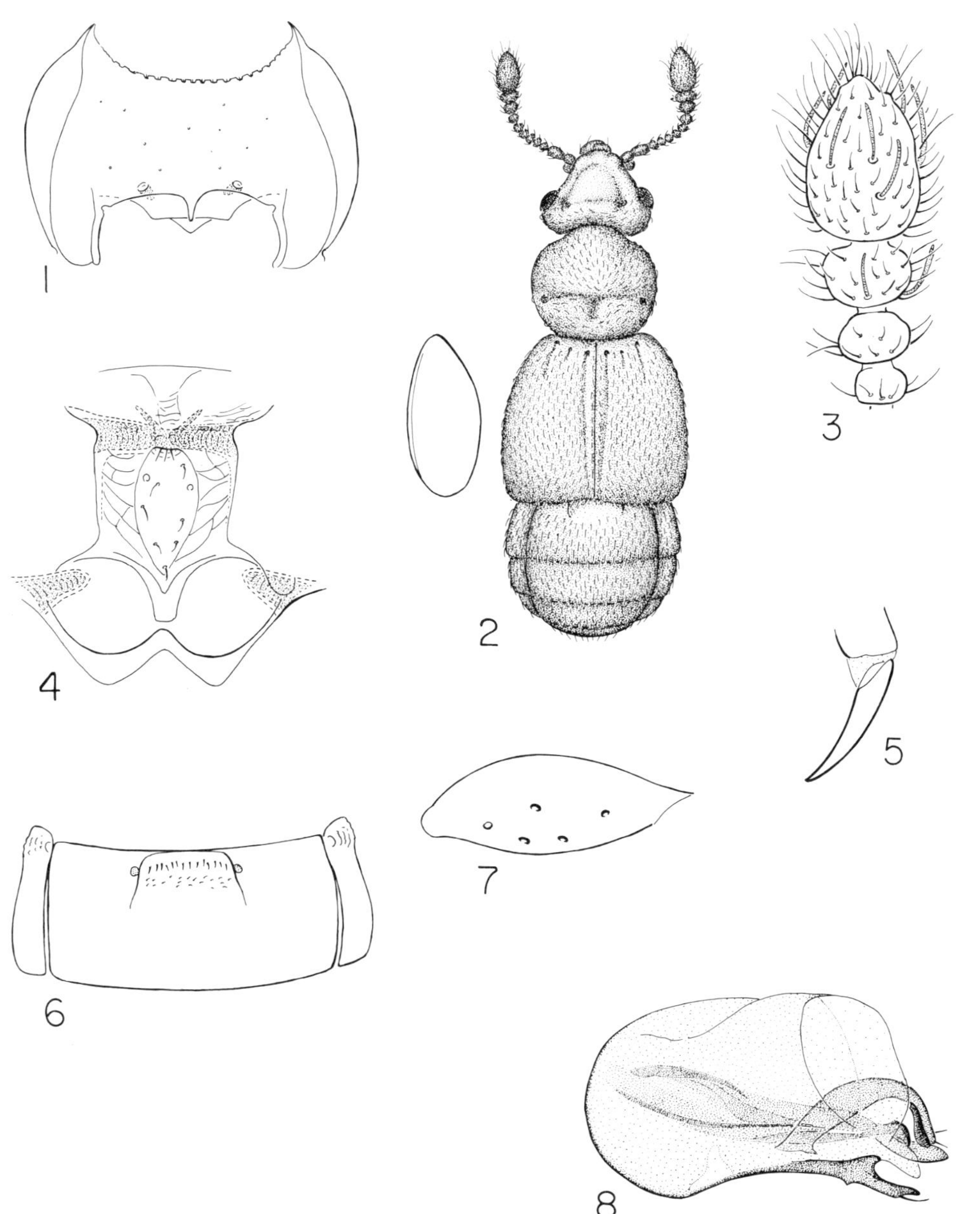

PLATE 8. *Trimioplectus obsoletus* Brendel. Group A. Figs. 1–5, 7 ♀; figs. 6, 8 ♂.

Fig. 1. Profemur.
Fig. 2. Dorsal aspect.
Fig. 3. Antennal segments IX to XI.
Fig. 4. Mesotarsal claws.
Fig. 5. Mesosternal area.
Fig. 6. Tergite I.
Fig. 7. Abdominal segments VIII and IX.
Fig. 8. Genitalia, dorsal aspect.

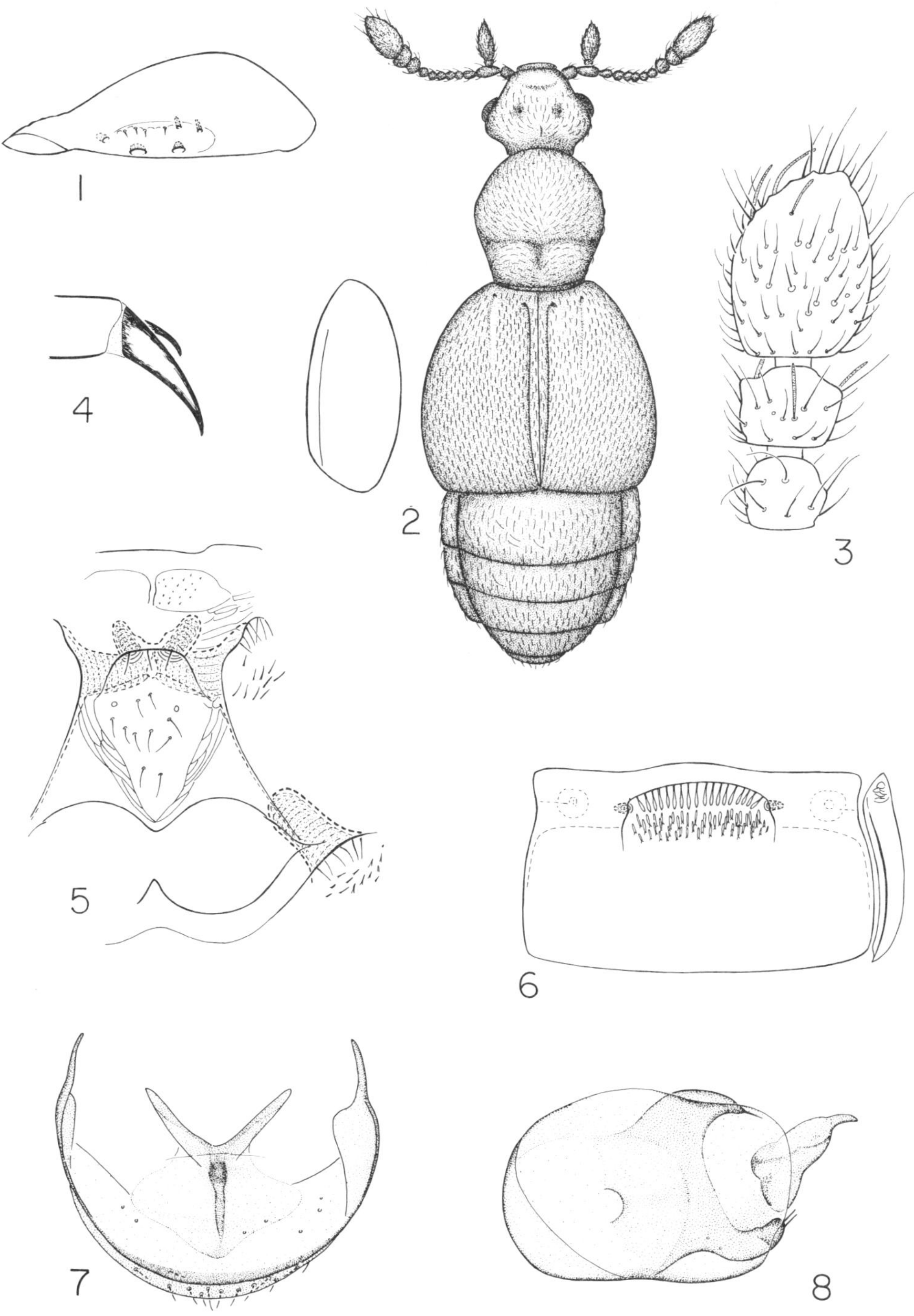

PLATE 9. *Pycnoplectus linearis* (LeConte). Group A. Figs. 3, 4, 6, 8 ♂; figs. 1, 2, 5, 7 ♀.

Fig. 1. Mesosternal area.
Fig. 2. Mesotarsal claw.
Fig. 3. Dorsal aspect.
Fig. 4. Antennal segments VIII to XI.
Fig. 5. Tergites I and II.
Fig. 6. Sternite II.
Fig. 7. Abdominal segments VIII and IX.
Fig. 8. Genitalia, lateral view.

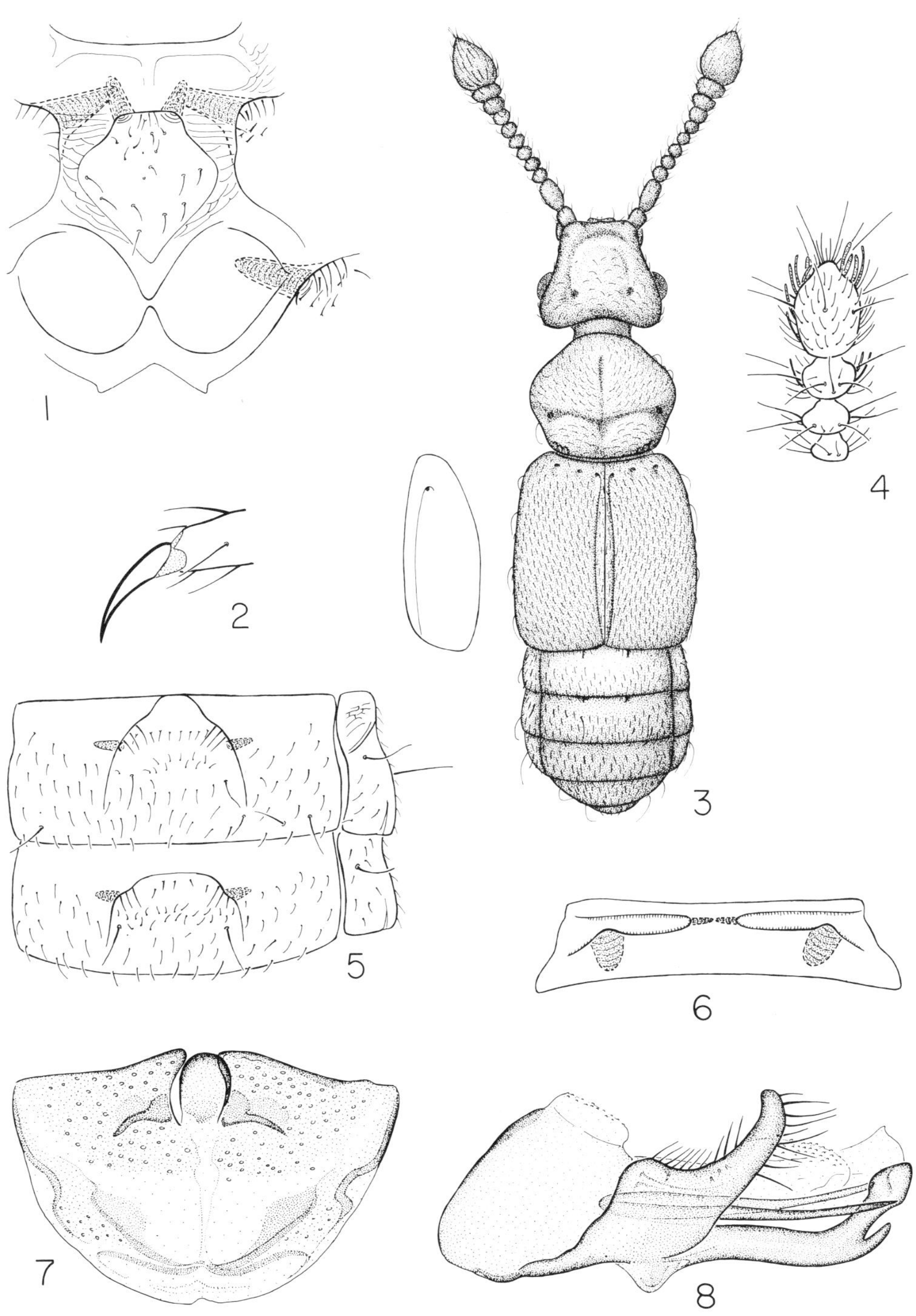

PLATE 10. *Acolonia cavicollis* (LeConte). Group A. Type ♀.

Fig. 1. Prosternum.
Fig. 2. Protarsal claw.
Fig. 3. Mesotarsal claw.
Fig. 4. Antennal segments VIII to XI.
Fig. 5. Mesosternal area.
Fig. 6. Mesotrochanter.
Fig. 7. Tergites I and II.
Fig. 8. Genitalia, dorsal aspect.

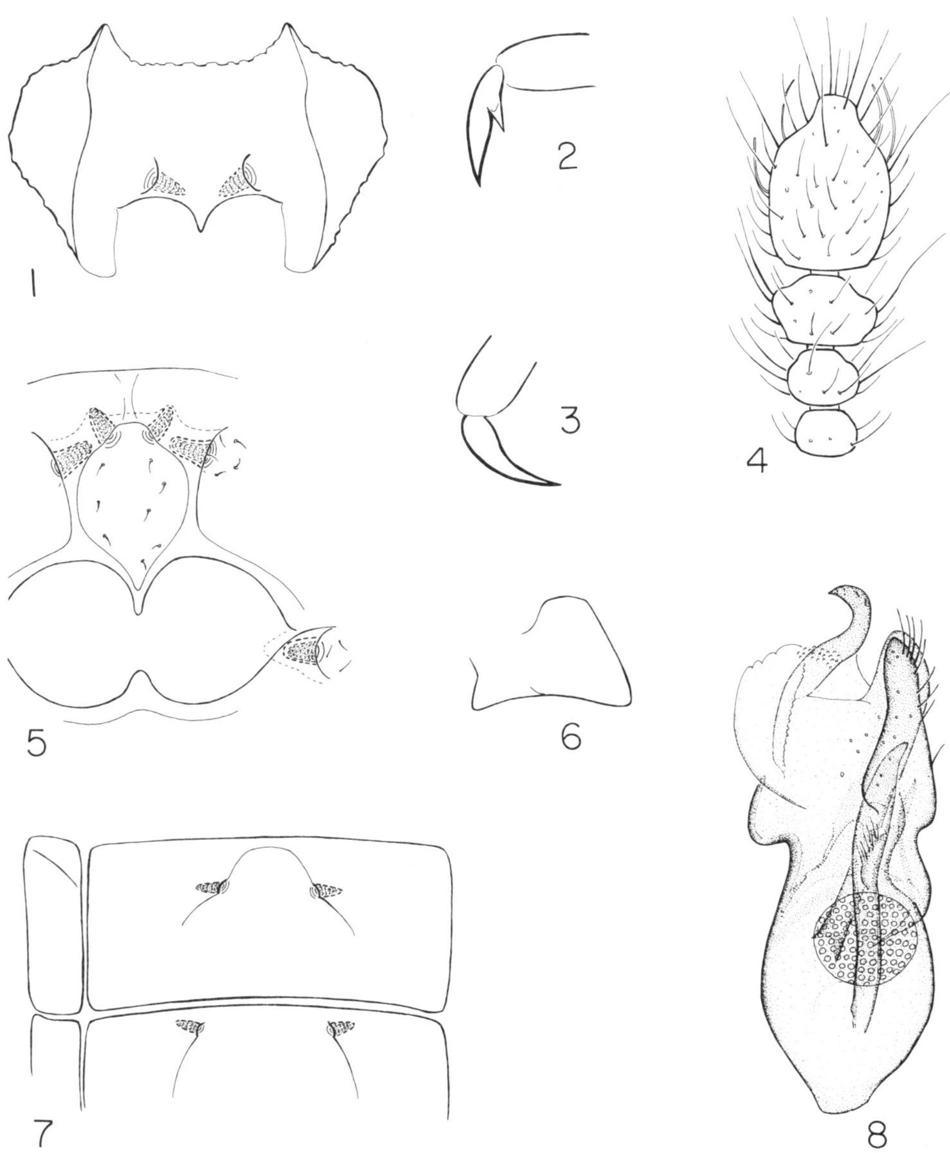

PLATE 11. *Barroeuplectoides zeteki* Park. Group A. Type ♀.

Fig. 1. Postantennal foveae.
Fig. 2. Prosternal area.
Fig. 3. Abdominal tergites I to V.
Fig. 4. Dorsal aspect.
Fig. 5. Antennal segments VIII to XI.
Fig. 6. Mesosternal area.
Fig. 7. Mesotarsi and claw.
Fig. 8. Coxal line of sternite I.

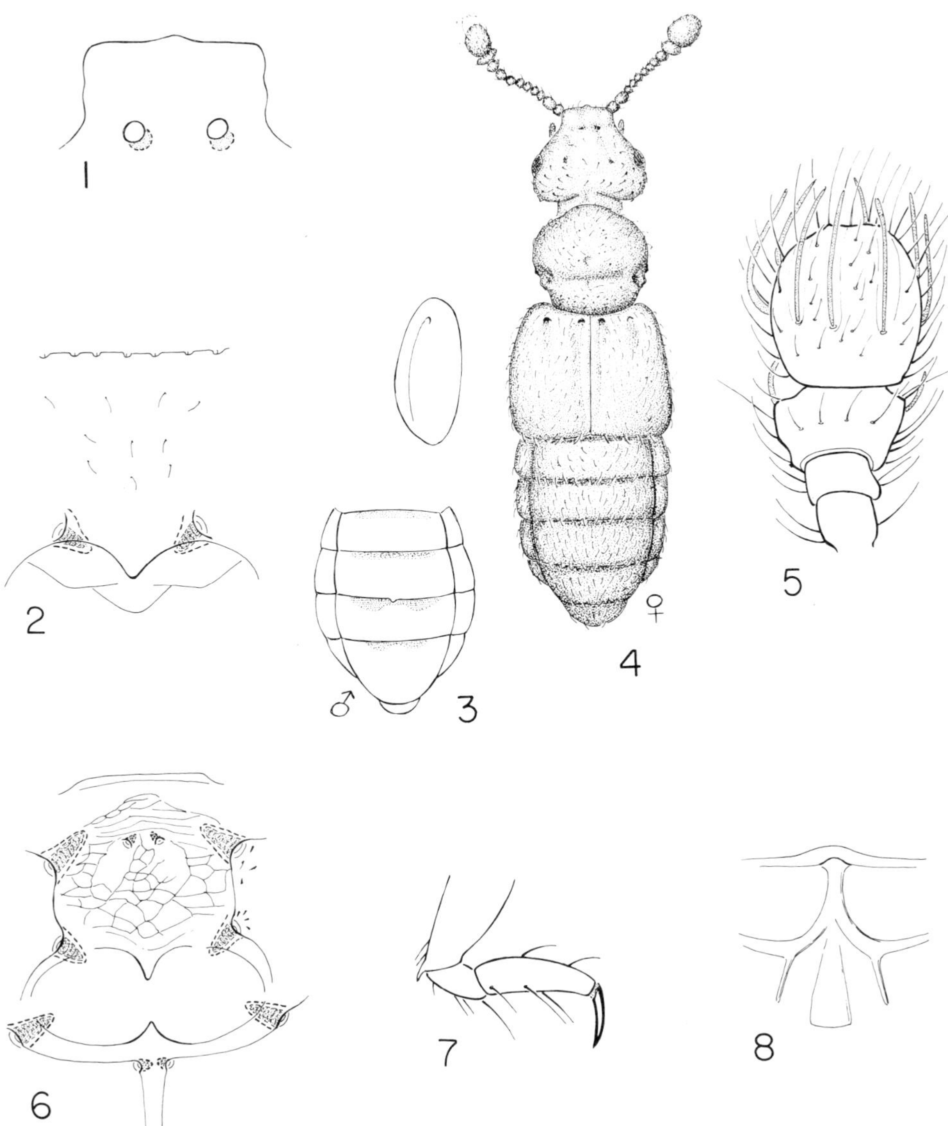

PLATE 12. *Tuberoplectus plaumanni* Park. Group A. Type ♀.

Fig. 1. Anterodorsum of head and postantennal foveae.
Fig. 2. Dorsal aspect.
Fig. 3. Antennal segments VIII to XI.
Fig. 4. Prosternal area.
Fig. 5. Protarsal claw.
Fig. 6. Coxal lines of sternite I.
Fig. 7. Mesosternal area.
Fig. 8. Basal depression of tergite I.

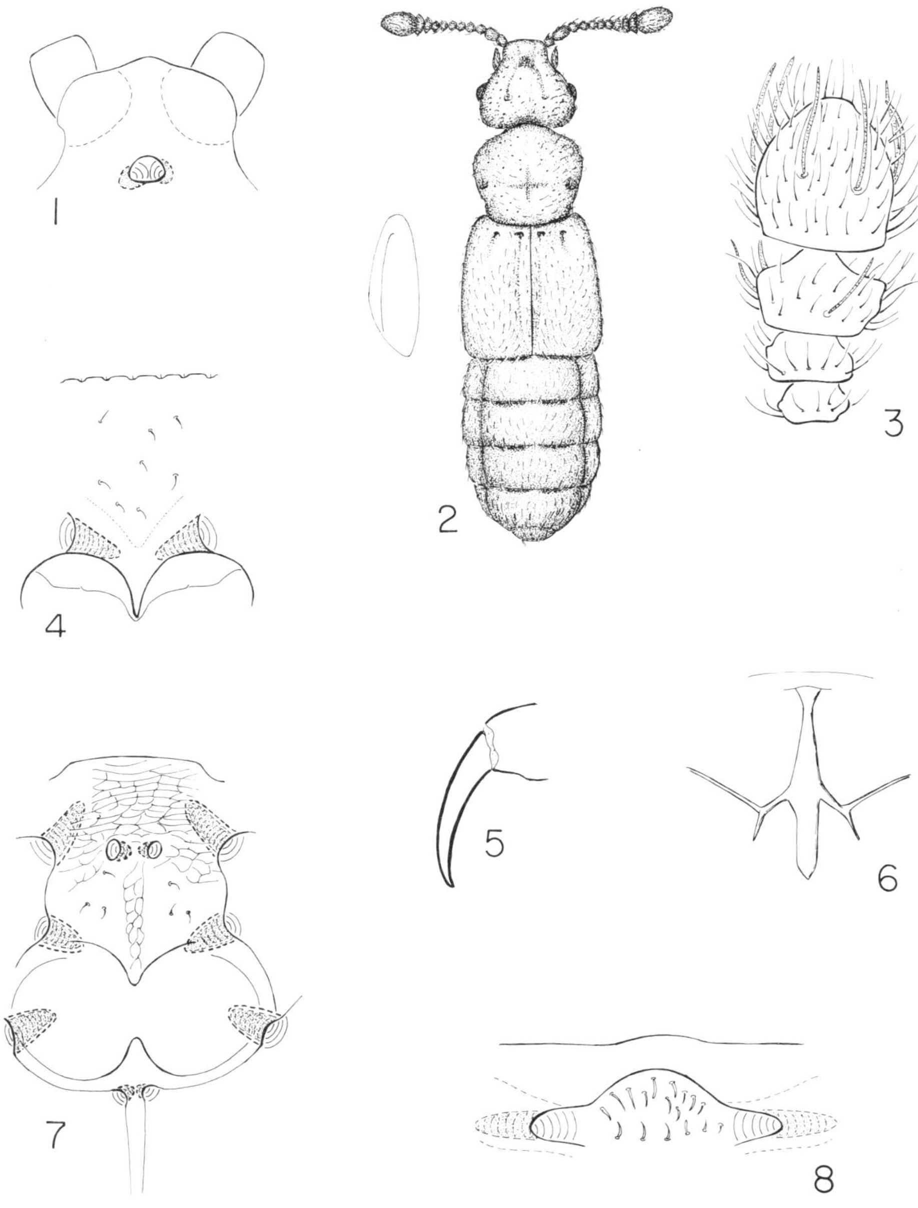

PLATE 13. *Oropus umbraticus* Schuster and Grigarick. Group A. Figs. 1, 2, 6, 7 ♂.
Oropus castaneus Casey. Fig. 3 ♂; fig. 12 ♀. *Oropus macneilli*
Schuster and Grigarick. Figs. 4, 5, 8, 9 ♂. *Oropus sinifundus*
Schuster and Grigarick. Figs. 10, 11 ♂.

Fig. 1. Ventral aspect of head.
Fig. 2. Mesosternal area.
Fig. 3. Dorsal aspect.
Fig. 4. Antennal segments VIII to XI.
Fig. 5. Profemur.
Fig. 6. Mesotarsal claws.
Fig. 7. Maxillary palp.
Fig. 8. Tergite I.
Fig. 9. Tergite V.
Fig. 10. Genitalia, dorsal aspect.
Fig. 11. Pronotum.
Fig. 12. Abdominal segments VIII and IX.

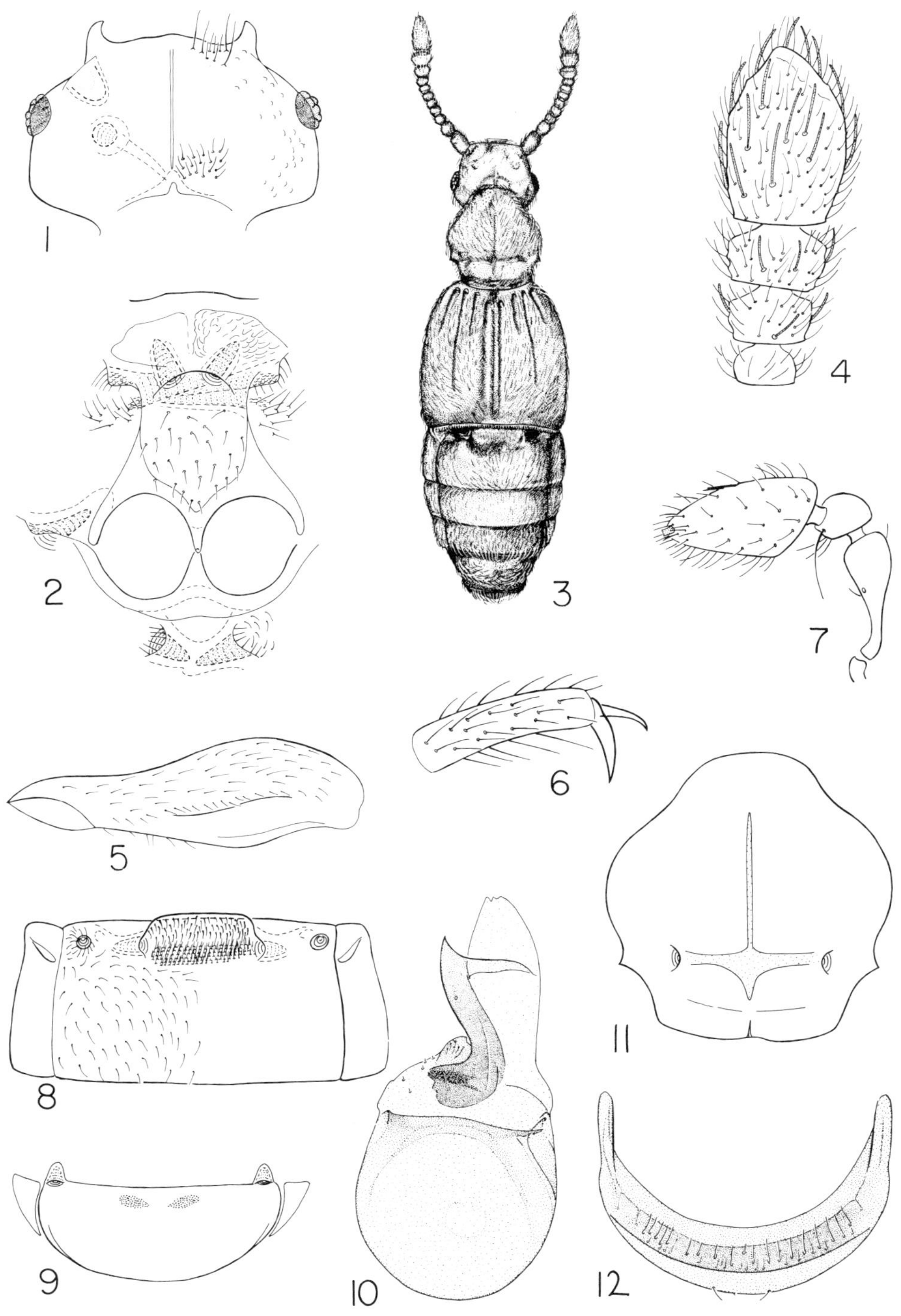

PLATE 14. *Rhexidius glareosus* Casey. Group A. Figs. 1, 2, 4, 6-9 ♂; figs. 3, 5, 10 ♀.

Fig. 1. Head, dorsal aspect.
Fig. 2. Antennal segments VIII to XI.
Fig. 3. Mesosternal area.
Fig. 4. Dorsal aspect.
Fig. 5. Metatarsal claws.
Fig. 6. Tergite I.
Fig. 7. Tergite V.
Fig. 8. Genitalia, dorsal aspect.
Fig. 9. Pronotum.
Fig. 10. Abdominal segments VIII and IX.

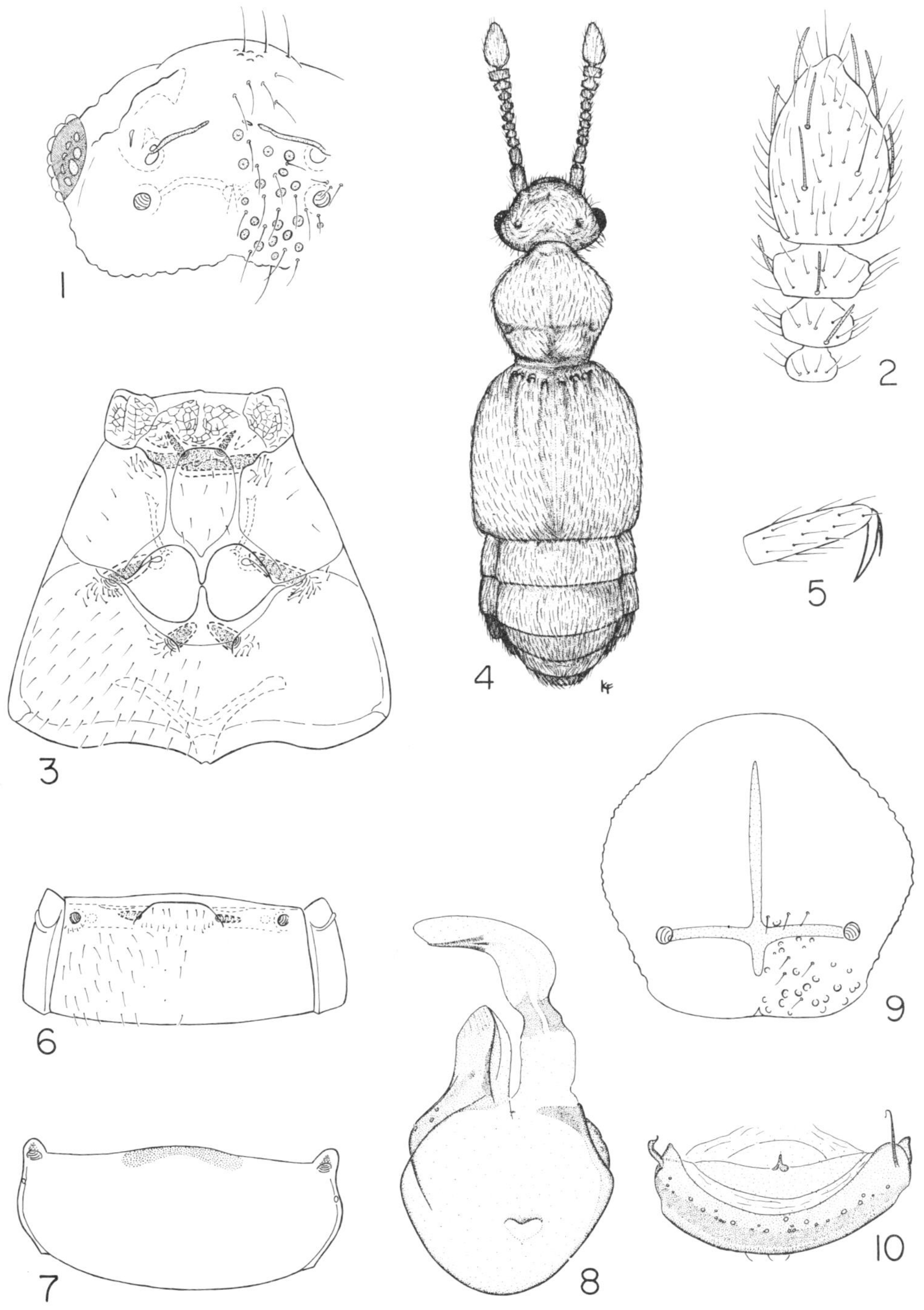

PLATE 15. *Fletcherexius macrodactylus* (Fletcher). Group A. Type ♀.

Fig. 1. Prosternum.
Fig. 2. Mesosternal area.
Fig. 3. Dorsal aspect.
Fig. 4. Antennal segments IX to XI.
Fig. 5. Metatarsal claw.
Fig. 6. Tergite I.
Fig. 7. Profemur.

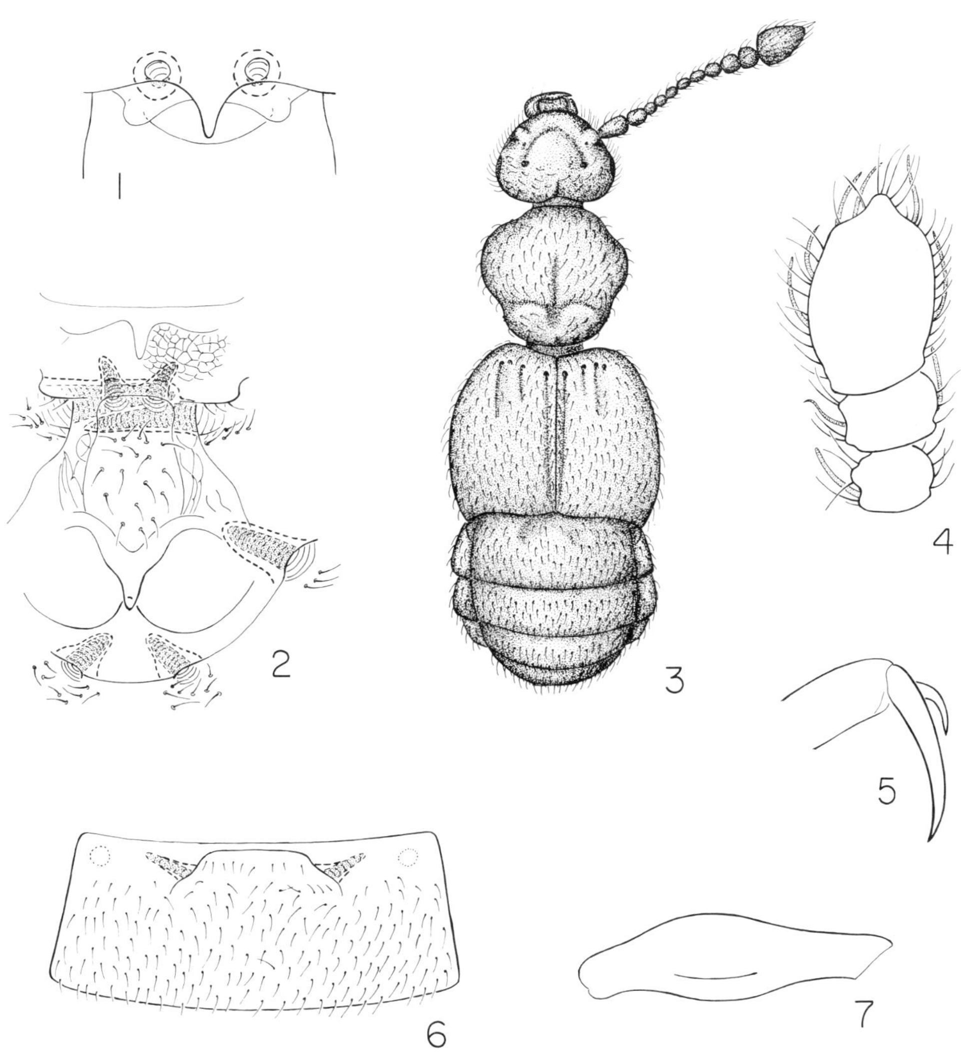

PLATE 16. *Hexirhexius canaliculatus* (LeConte). Group A. Homotype ♂, ♀. Figs. 1–5, 7 ♂; fig. 6 ♀.

Fig. 1. Head, dorsal aspect.
Fig. 2. Antennal segments IX to XI.
Fig. 3. Mesosternal area.
Fig. 4. Dorsal aspect.
Fig. 5. Metatarsal claws.
Fig. 6. Abdominal segments VIII and IX.
Fig. 7. Genitalia, dorsal aspect.

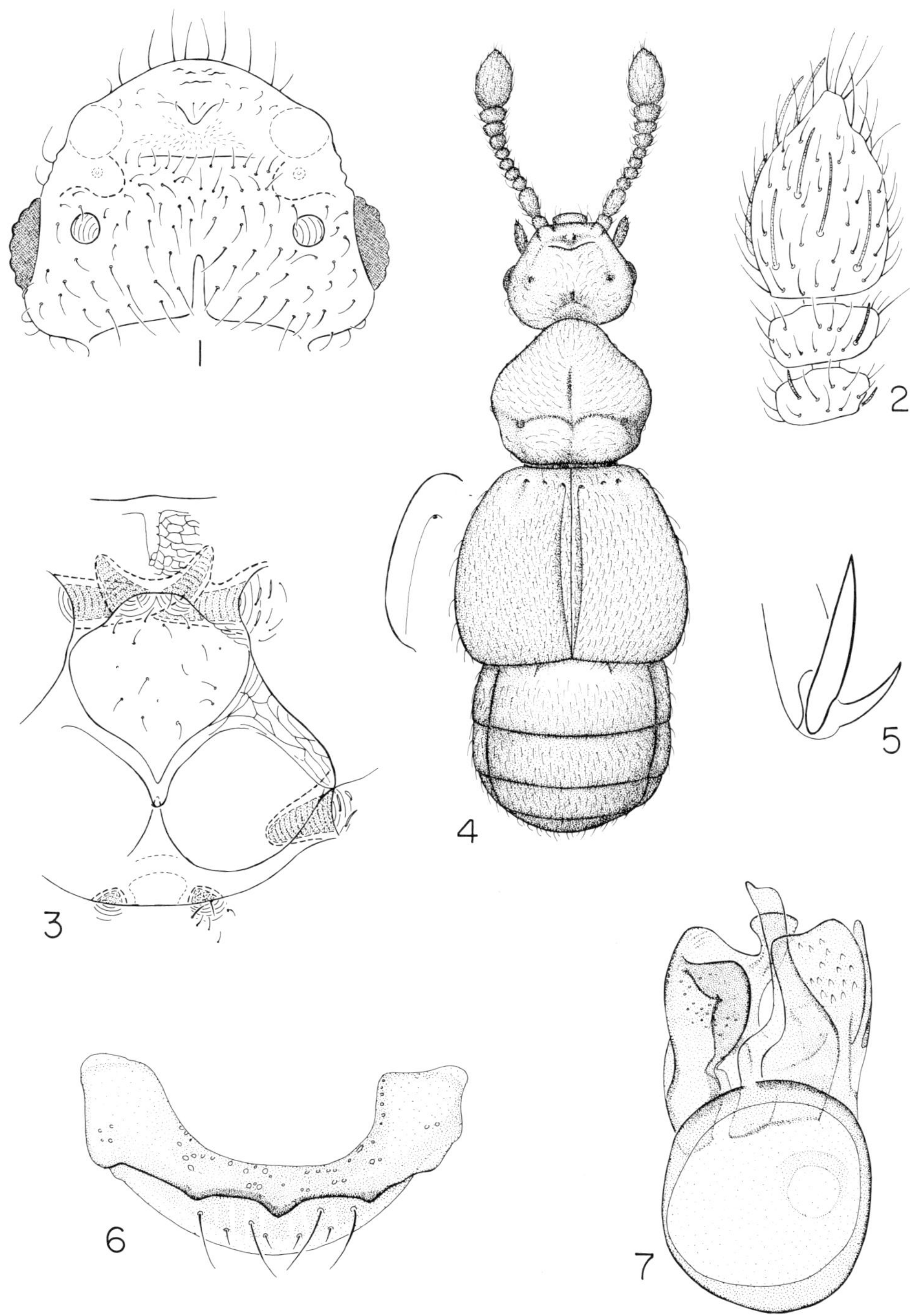

PLATE 17. *Euboarhexius sinus* Grigarick and Schuster. Group A. Type ♂.

Fig. 1. Antenna.
Fig. 2. Mesosternal area.
Fig. 3. Mesotarsal claws.
Fig. 4. Prosternum.
Fig. 5. Genitalia, dorsal aspect.

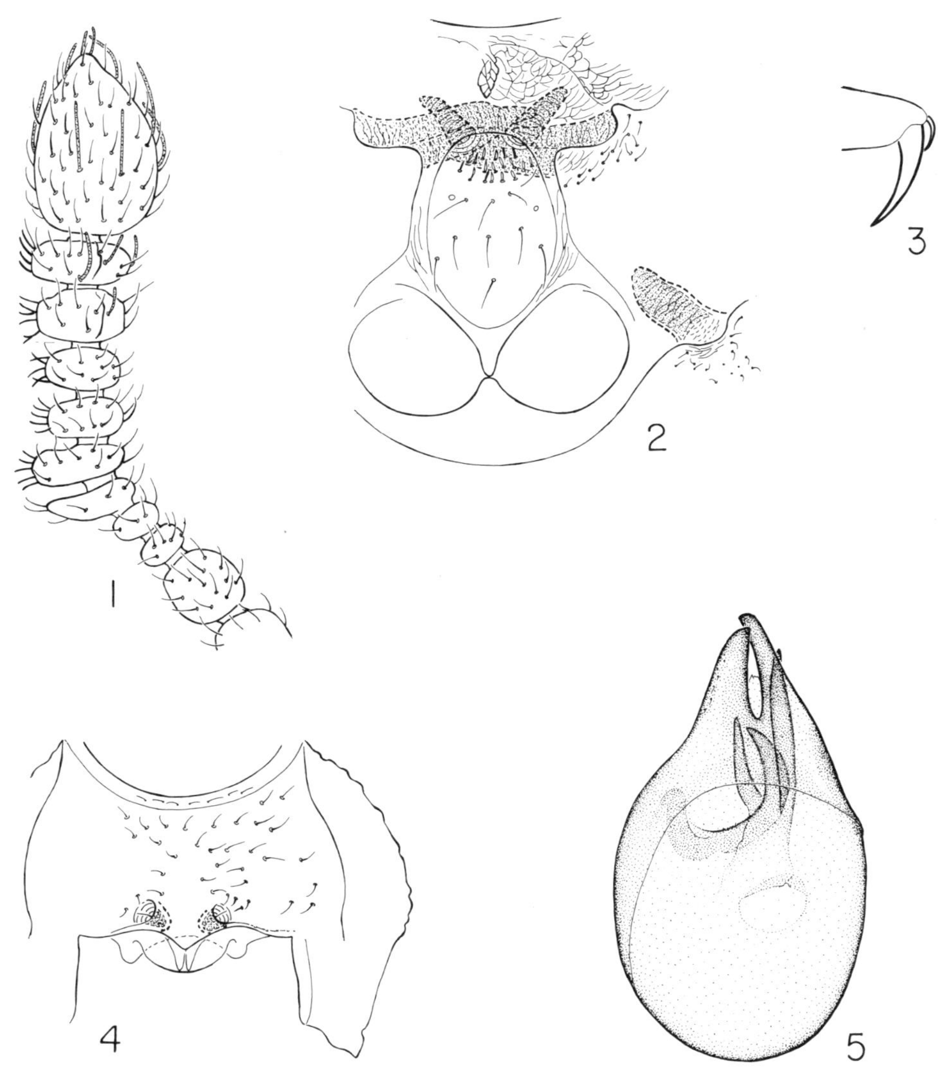

PLATE 18. *Rhexius sharpi* Park. Group A. Figs. 1-6 ♀; fig. 7 ♂.

Fig. 1. Mesosternal area.
Fig. 2. Dorsal aspect.
Fig. 3. Antennal segments VIII to XI.
Fig. 4. Abdominal segments VIII and IX.
Fig. 5. Metatarsal claws.
Fig. 6. Tergite I.
Fig. 7. Genitalia, dorsal aspect.

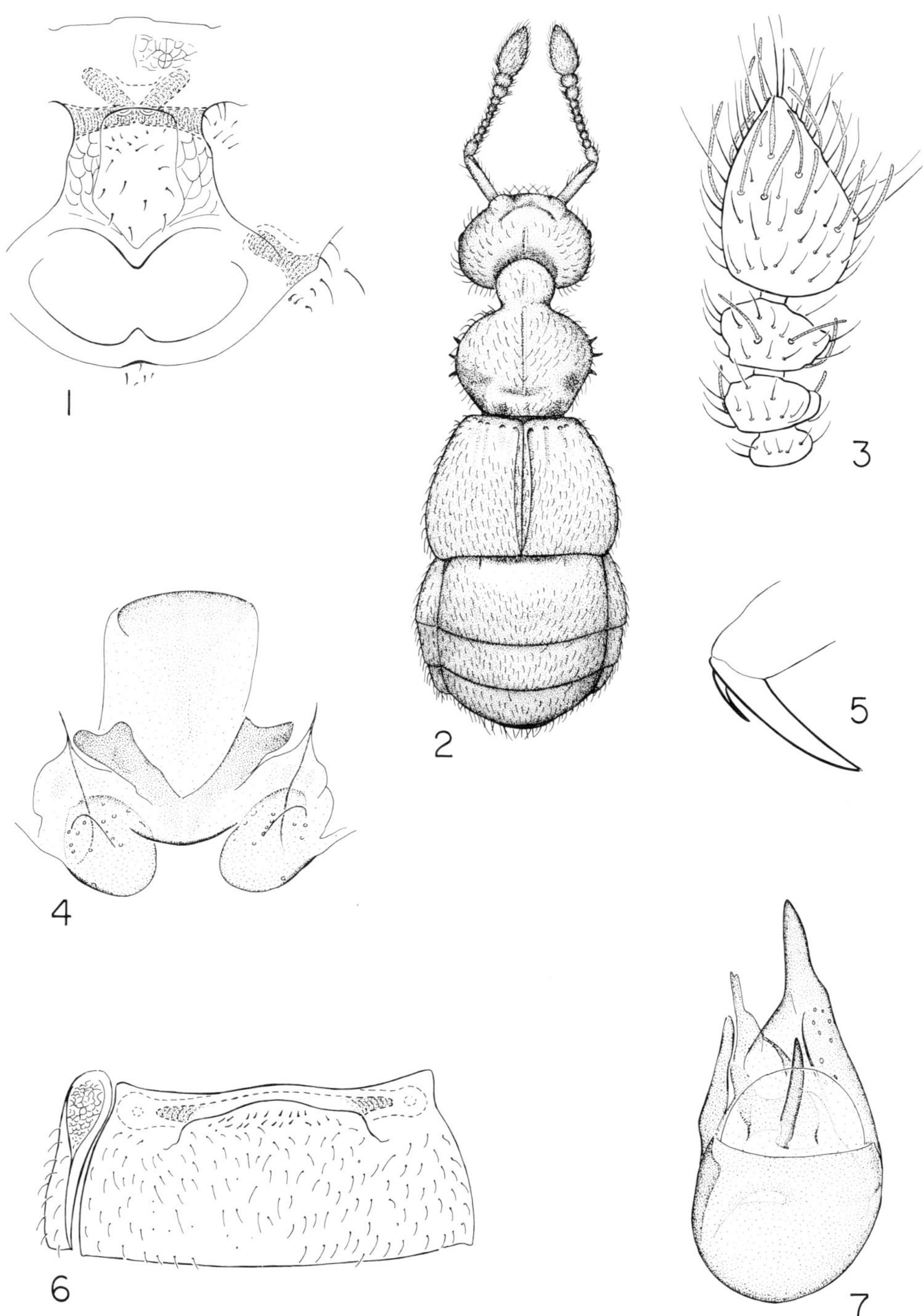

PLATE 19. *Eurhexius tropicus* Park. Group A. Paratype ♂, ♀. Figs. 1, 2, 4–8 ♂; fig. 3 ♀.

Fig. 1. Ventral aspect of head.
Fig. 2. Prosternal area.
Fig. 3. Dorsal aspect.
Fig. 4. Antennal segments IX to XI.
Fig. 5. Mesotarsal claws.
Fig. 6. Pores of profemur.
Fig. 7. Mesosternal area.
Fig. 8. Genitalia, dorsal aspect.

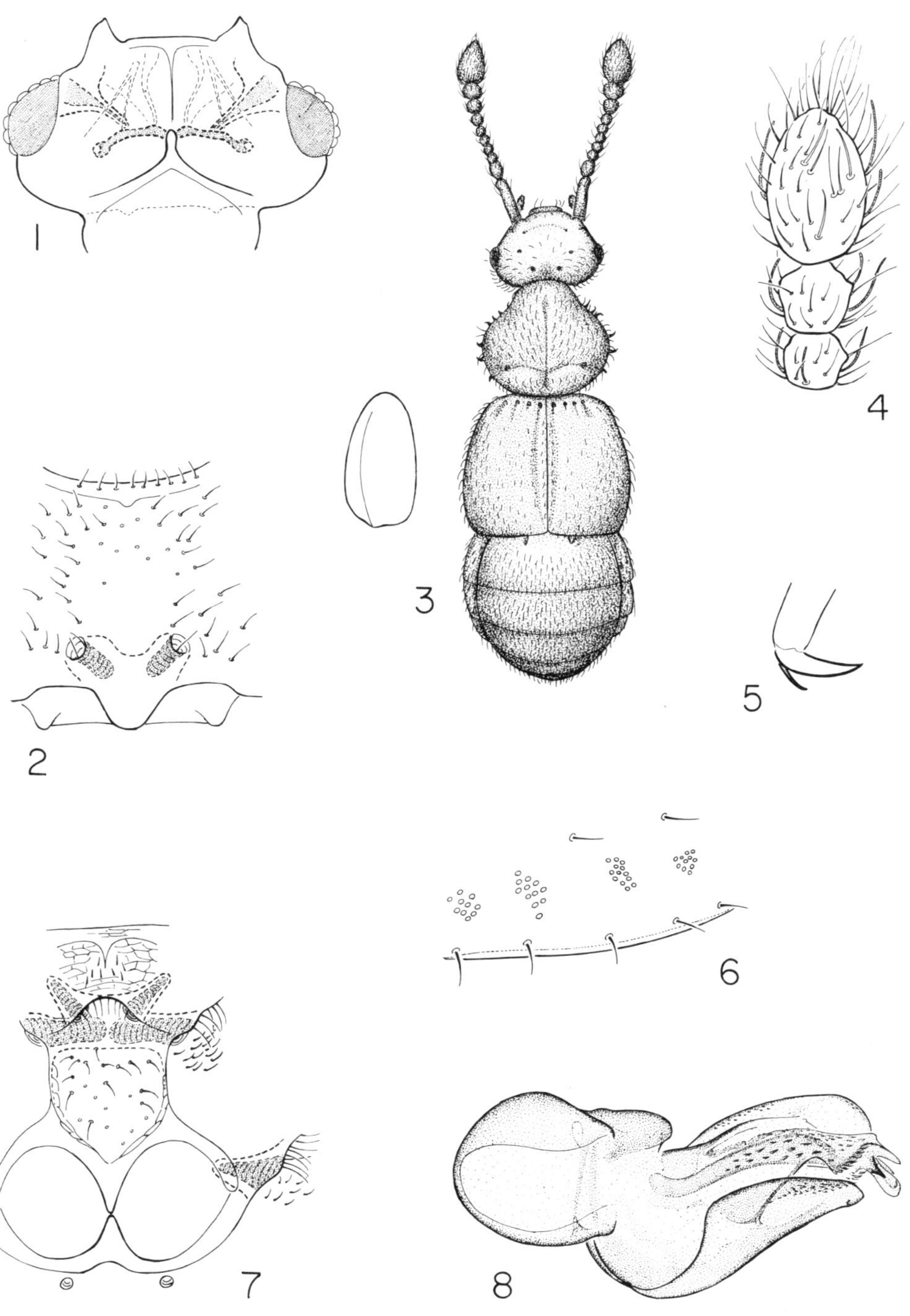

PLATE 20. *Euplectus confluens* LeConte. Group A. ♂. Figs. 1, 2, 4–6. *Euplectus sp.*
♂. Fig. 3.

Fig. 1. Prosternum.
Fig. 2. Tergites I and II.
Fig. 3. Dorsal aspect.
Fig. 4. Sternites IV to VI.
Fig. 5. Genitalia, lateral aspect.
Fig. 6. Mesosternal area.

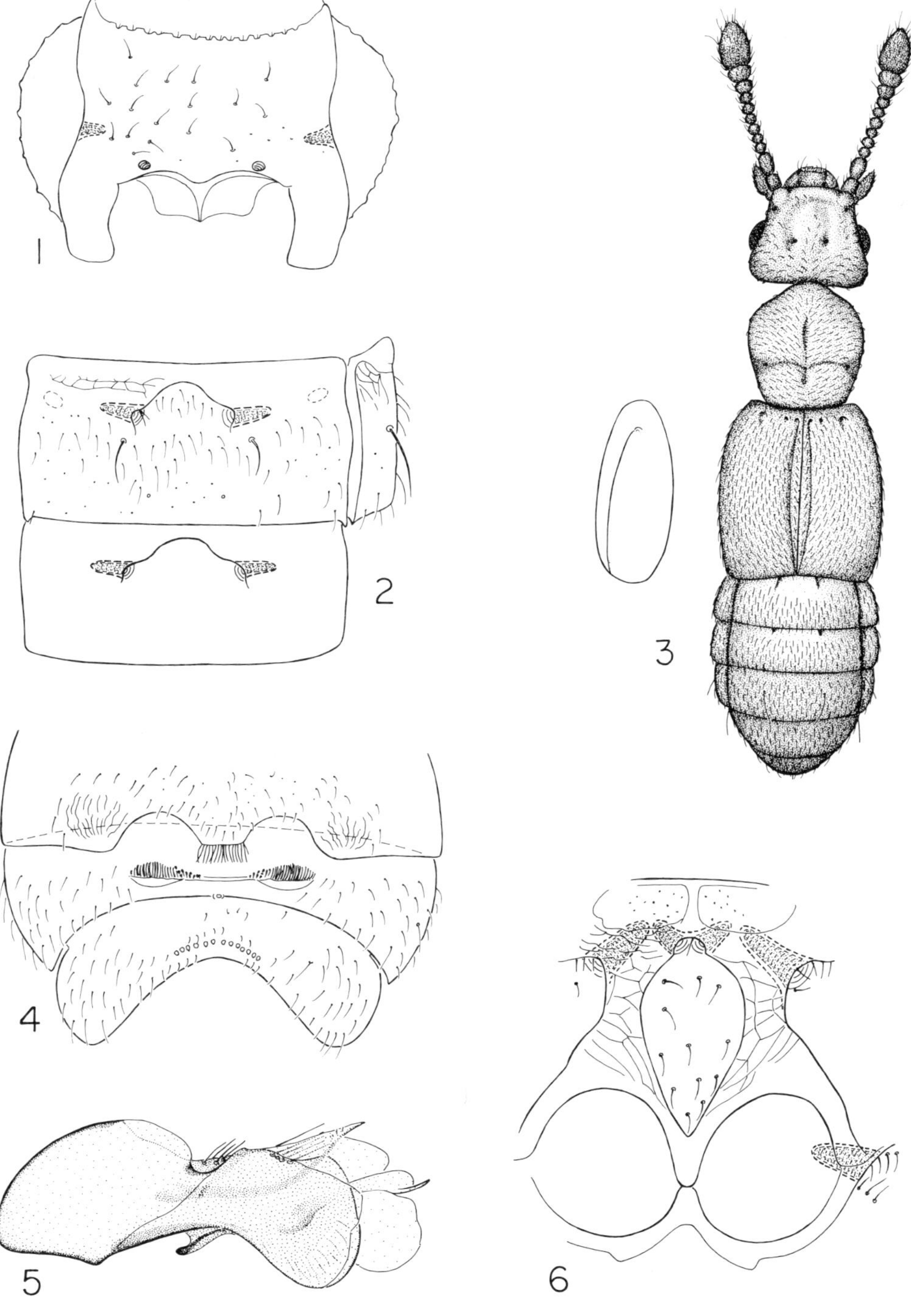

PLATE 21. *Rhexinia angulata* Raffray. Group A. ♂.

Fig. 1. Head, dorsal, foveae, and apodemes.
Fig. 2. Prosternal area.
Fig. 3. Dorsal aspect.
Fig. 4. Antennal segments IX to XI.
Fig. 5. Mesotarsal claws.
Fig. 6. Mesosternal area.
Fig. 7. Profemur.
Fig. 8. Genitalia, dorsal aspect.

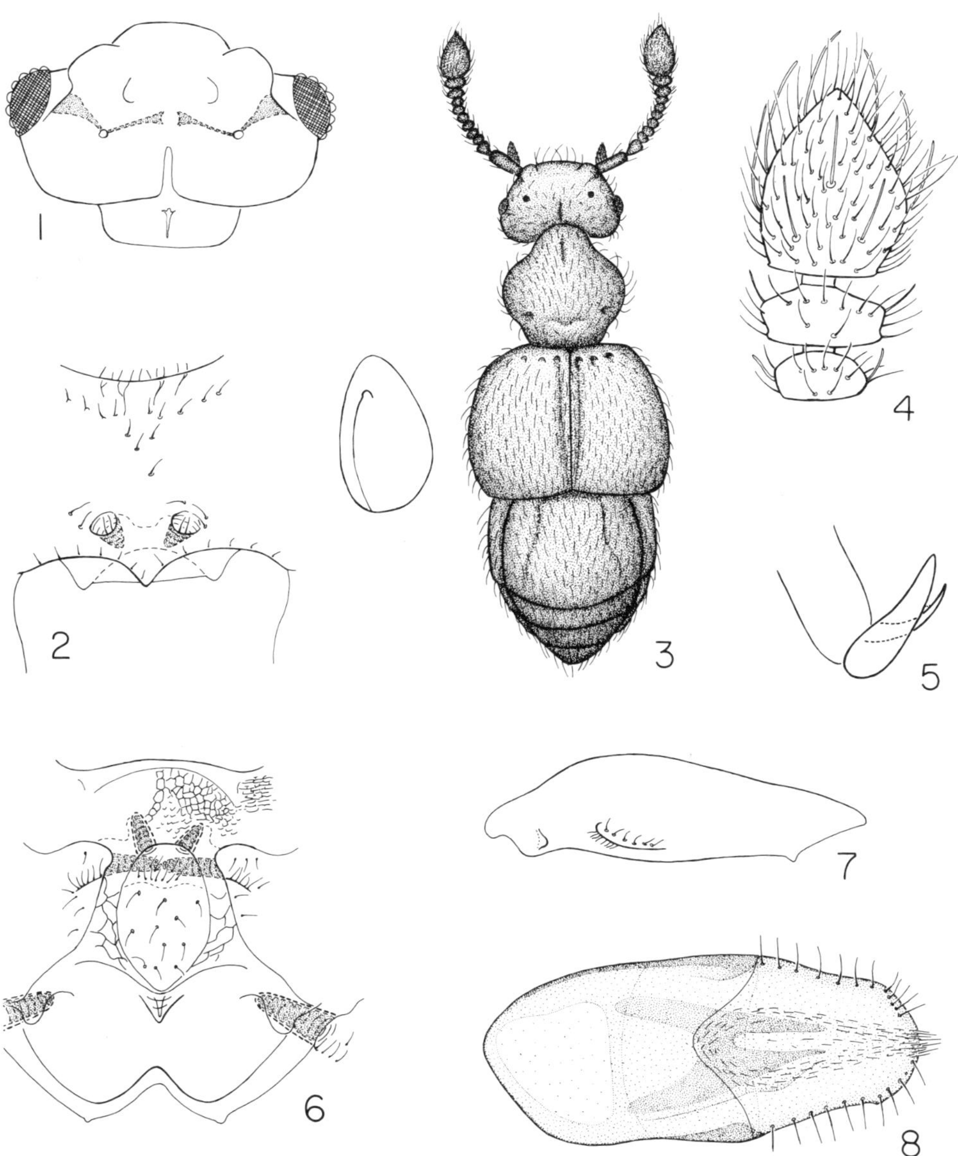

PLATE 22. *Rhexiola plana* Park. Group A. Paratype ♂.

Fig. 1. Mesosternal area.
Fig. 2. Dorsal aspect.
Fig. 3. Antennal segments VIII to XI.
Fig. 4. Protarsal claws.
Fig. 5. Genitalia, dorsal aspect.

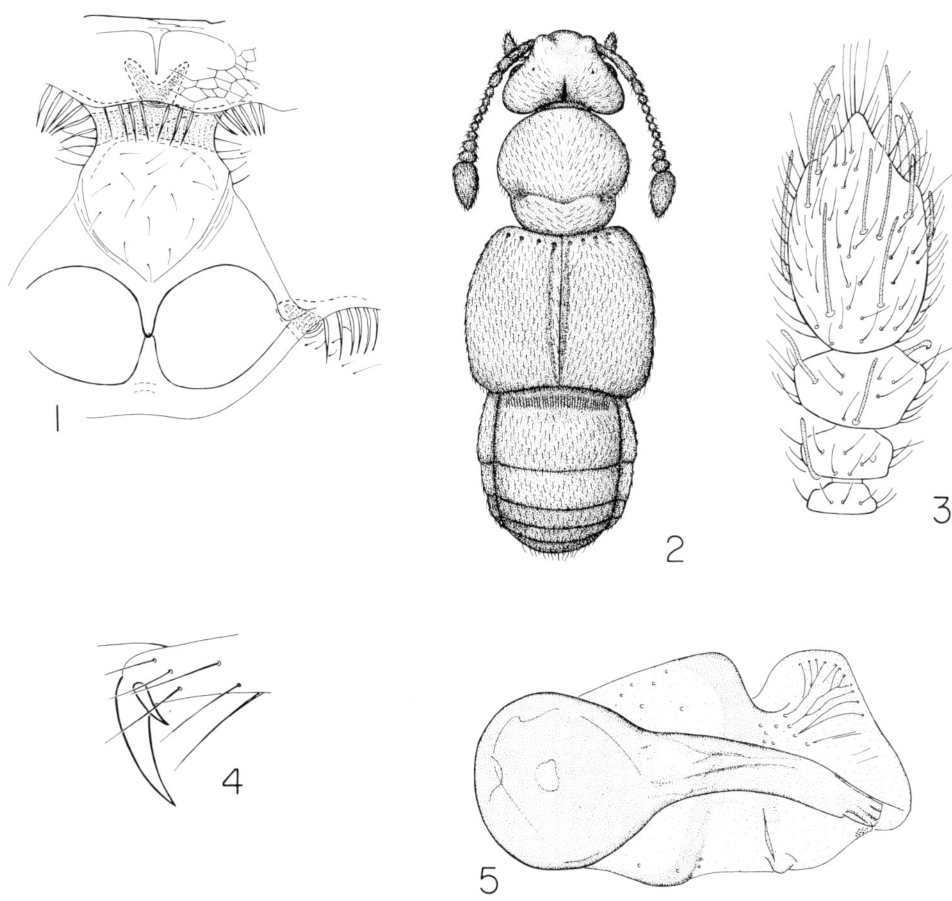

PLATE 23. *Hatchia carinata* Park and Wagner. Group A. Type ♂, paratype ♀. Figs. 3-6 ♂; figs. 1, 2, 7 ♀.

Fig. 1. Ventral aspect of head.
Fig. 2. Prosternum.
Fig. 3. Dorsal aspect.
Fig. 4. Mesosternal area.
Fig. 5. Mesotarsal claw.
Fig. 6. Antennal segments VIII to XI.
Fig. 7. Abdominal segments VIII and IX.

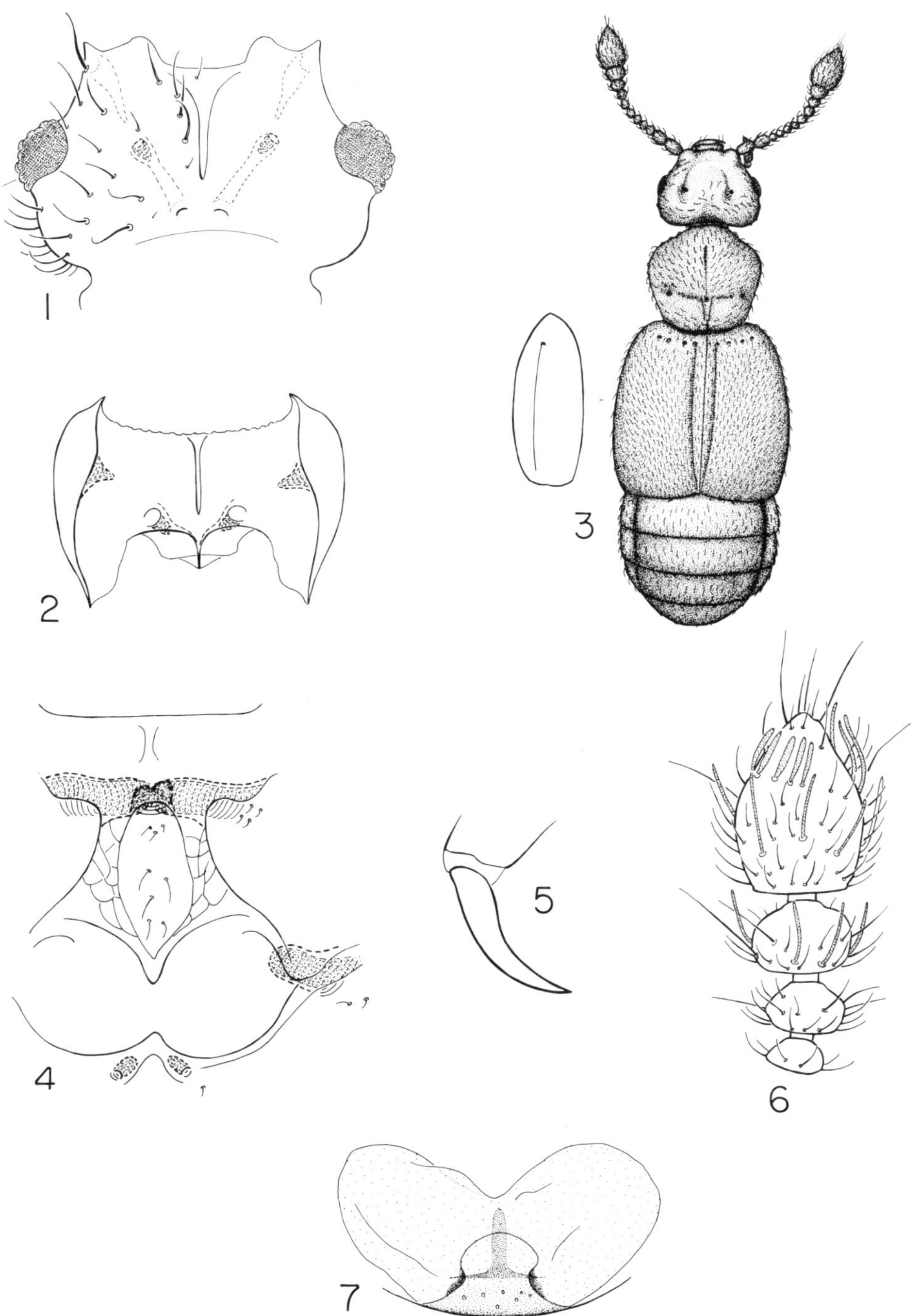

PLATE 24. *Bibloporus bicanalis* Casey. Group A. Cotype ♂. Figs. 1-7,9. *Bibloporus bicolor* (Denny). Fig. 8 ♀.

Fig. 1. Mesosternal area.
Fig. 2. Metatarsal claws.
Fig. 3. Dorsal aspect.
Fig. 4. Antennal segments VIII to XI.
Fig. 5. Setae on venter of head.
Fig. 6. Mesofemur and mesotibia.
Fig. 7. Tergite I.
Fig. 8. Abdominal segments VIII and IX.
Fig. 9. Genitalia, lateral aspect.

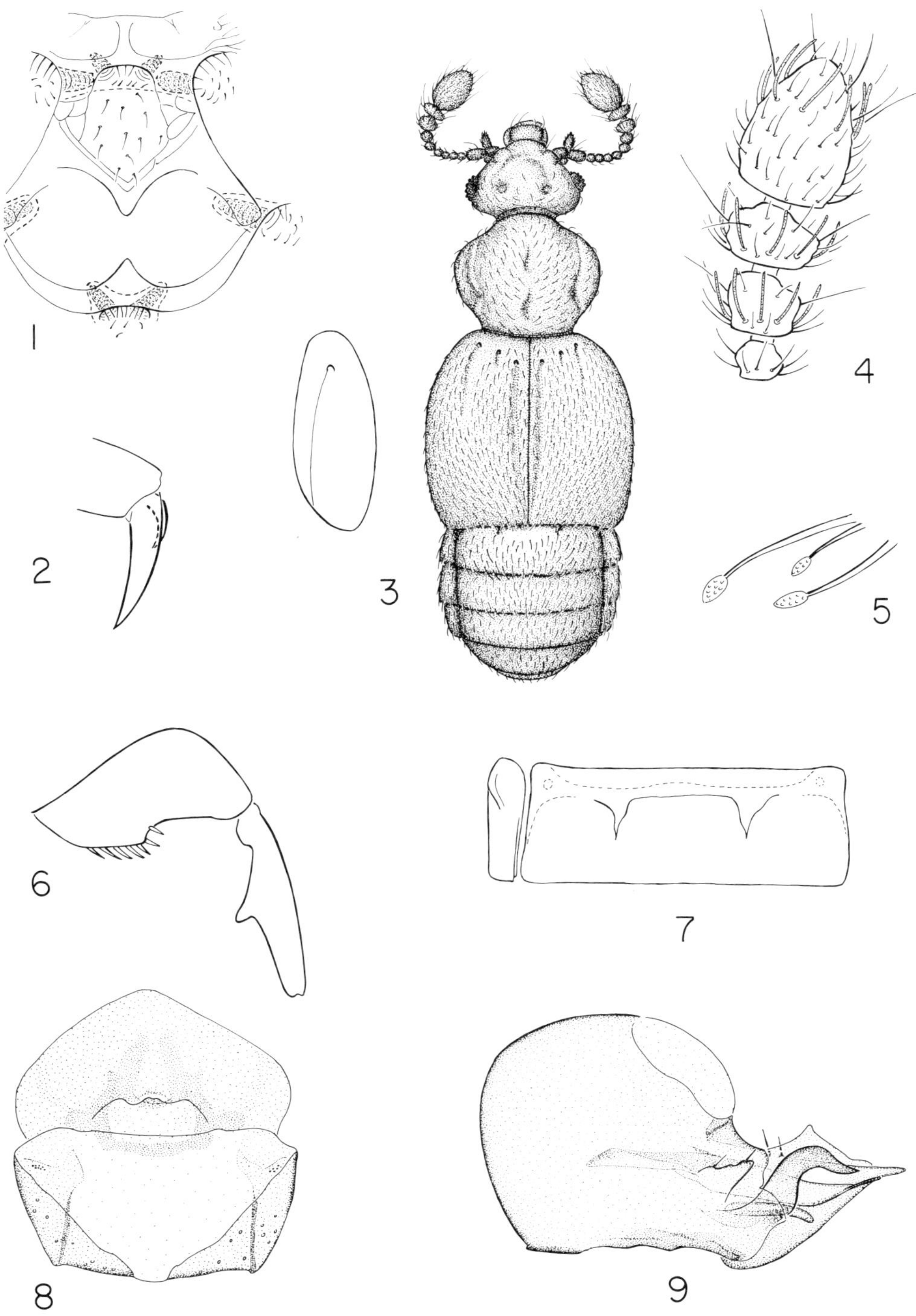

PLATE 25. *Aboeurhexius crenulatus* Park. Group A. Type ♀.

Fig. 1. Ventral aspect of head.
Fig. 2. Labium.
Fig. 3. Dorsal aspect.
Fig. 4. Antennal segments IX to XI.
Fig. 5. Mesosternal area.
Fig. 6. Metatarsal claws.
Fig. 7. "Maxillary brush."
Fig. 8. Tergite I.
Fig. 9. Sternite II.

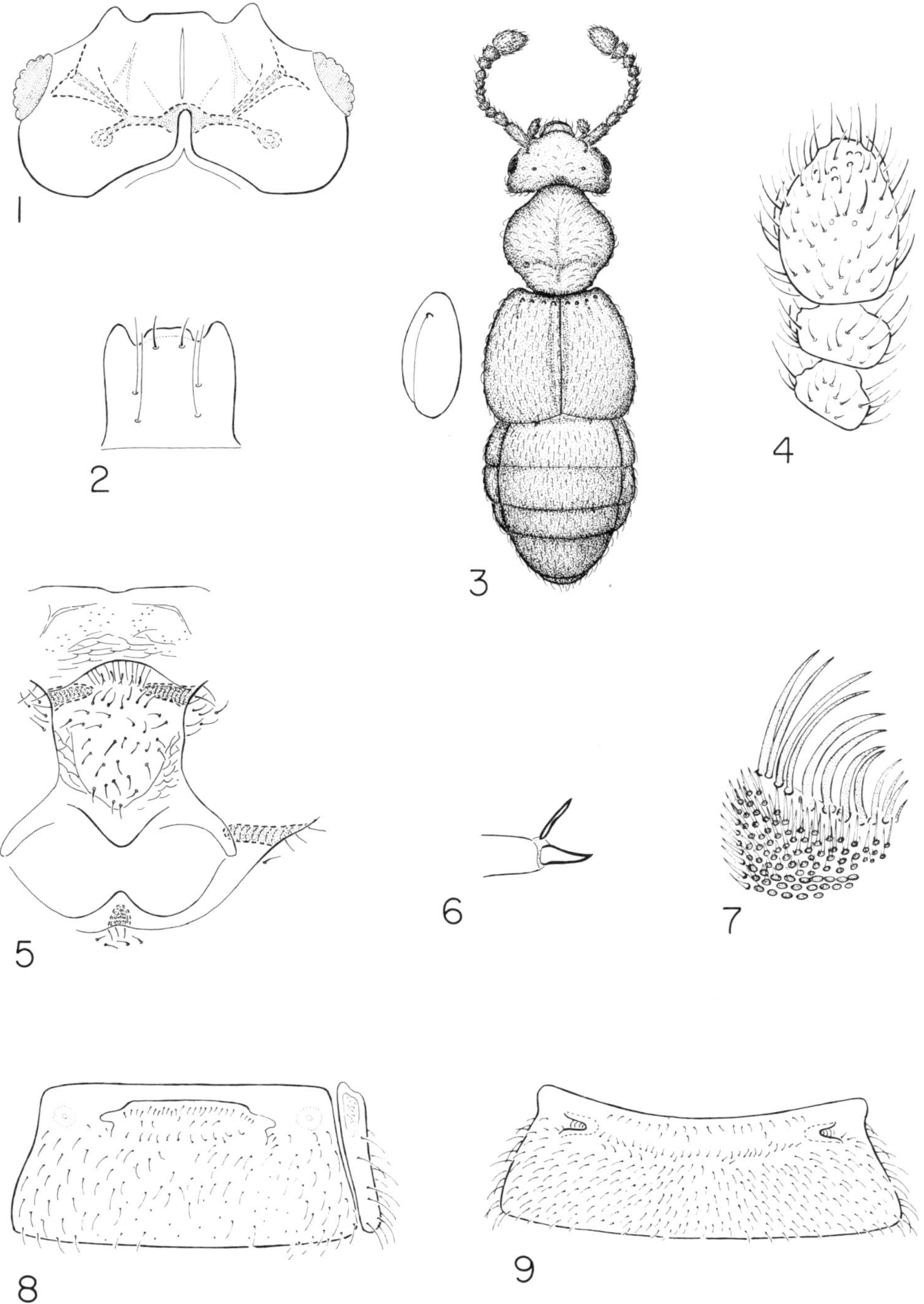

1
2
3
4
5
6
7
8
9

PLATE 26. *Morius occidens* Casey. Group B. Figs. 2, 6, 9 ♂; figs. 1, 3–5, 7, 8 ♀.

Fig. 1. Palpal segment IV.
Fig. 2. Prosternal area.
Fig. 3. Dorsal aspect.
Fig. 4. Antennal segments IX to XI.
Fig. 5. Mesosternal area.
Fig. 6. Mesotarsal claws.
Fig. 7. Tergite I.
Fig. 8. Abdominal segments VIII and IX.
Fig. 9. Genitalia, dorsal aspect.

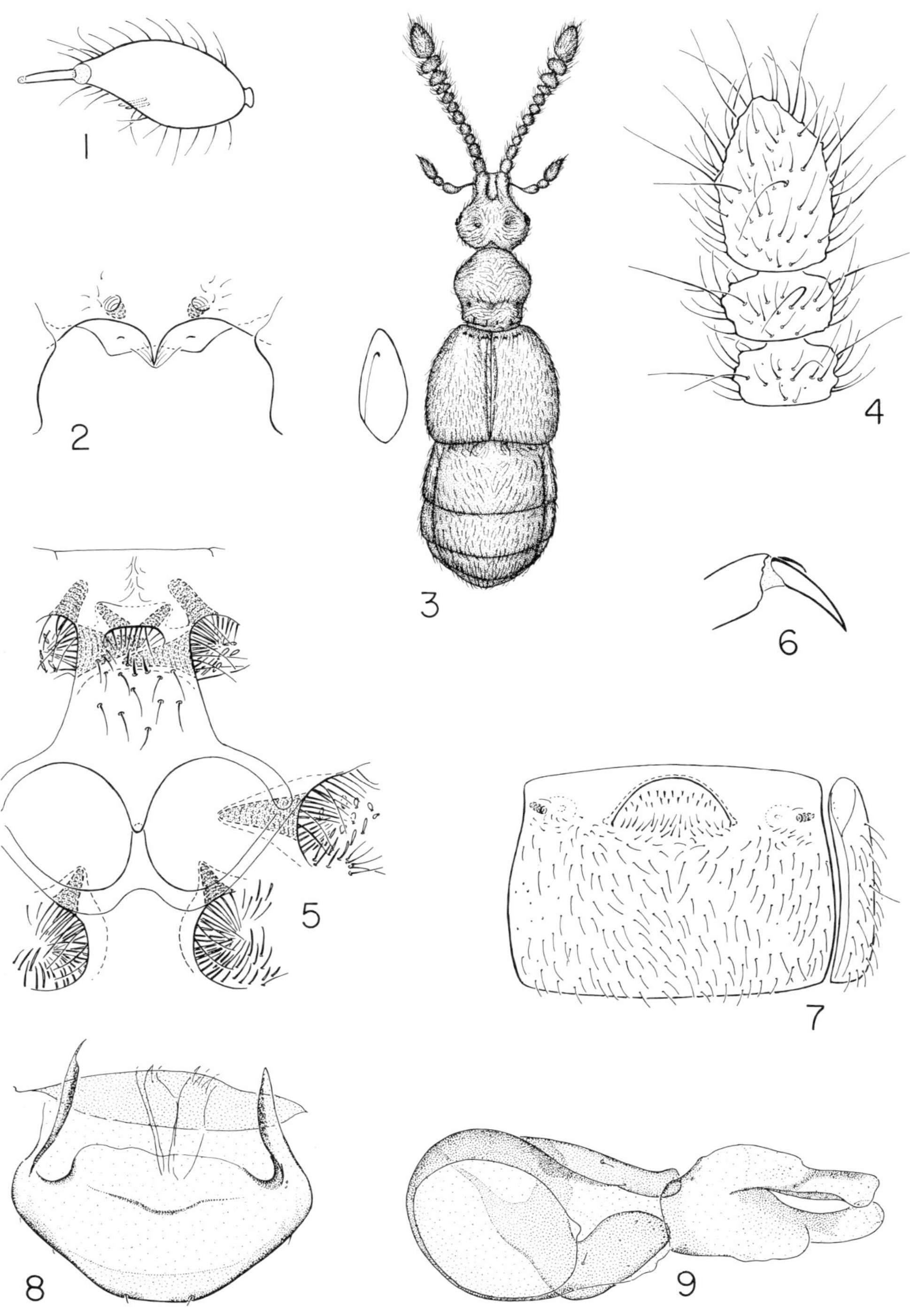

PLATE 27. *Bontomtes riparie* Grigarick and Schuster. Group B. Type ♂.

Fig. 1. Mesotarsal claws.
Fig. 2. Mesosternal area.
Fig. 3. Spur of protrochanter.
Fig. 4. Distribution of elytral foveae.
Fig. 5. Antennal segments VIII to XI.
Fig. 6. Prosternum.
Fig. 7. Male sex-limited character of sternite III.
Fig. 8. Tergite I.
Fig. 9. Genitalia, lateral aspect.

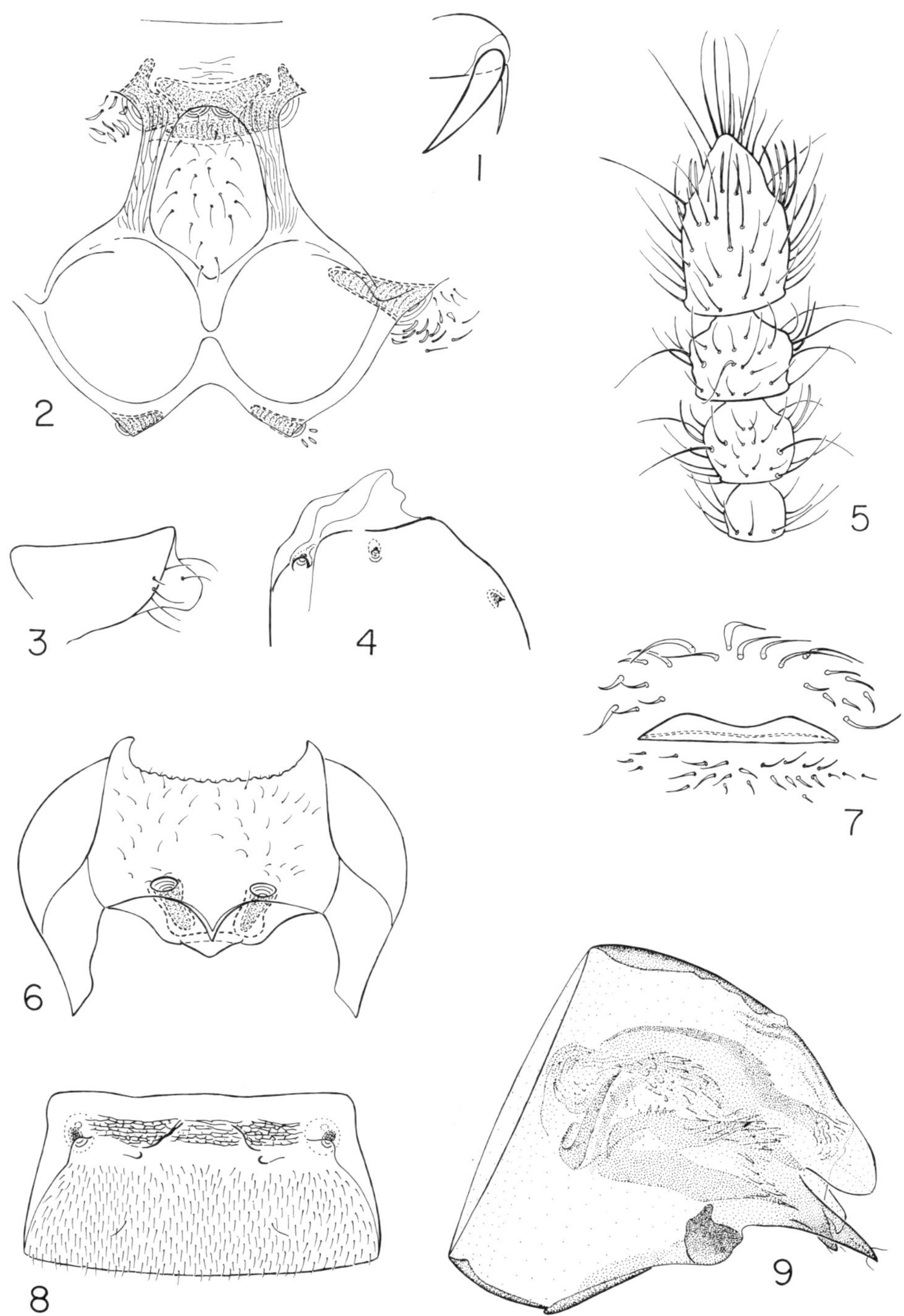

PLATE 28. *Foveoscapha terracola* Park and Wagner. Group B. Homotype ♂, ♀. Figs. 1-4, 6, 7, 9 ♂; figs. 5, 8 ♀.

Fig. 1. Distal segments of maxillary palp.
Fig. 2. Right anterior margin of sternite II.
Fig. 3. Dorsal aspect.
Fig. 4. Antennal segments VIII to XI.
Fig. 5. Metatarsal claws.
Fig. 6. Mesosternal area.
Fig. 7. Sex-limited character of sternite III.
Fig. 8. Abdominal segments VIII and IX.
Fig. 9. Genitalia, lateral aspect.

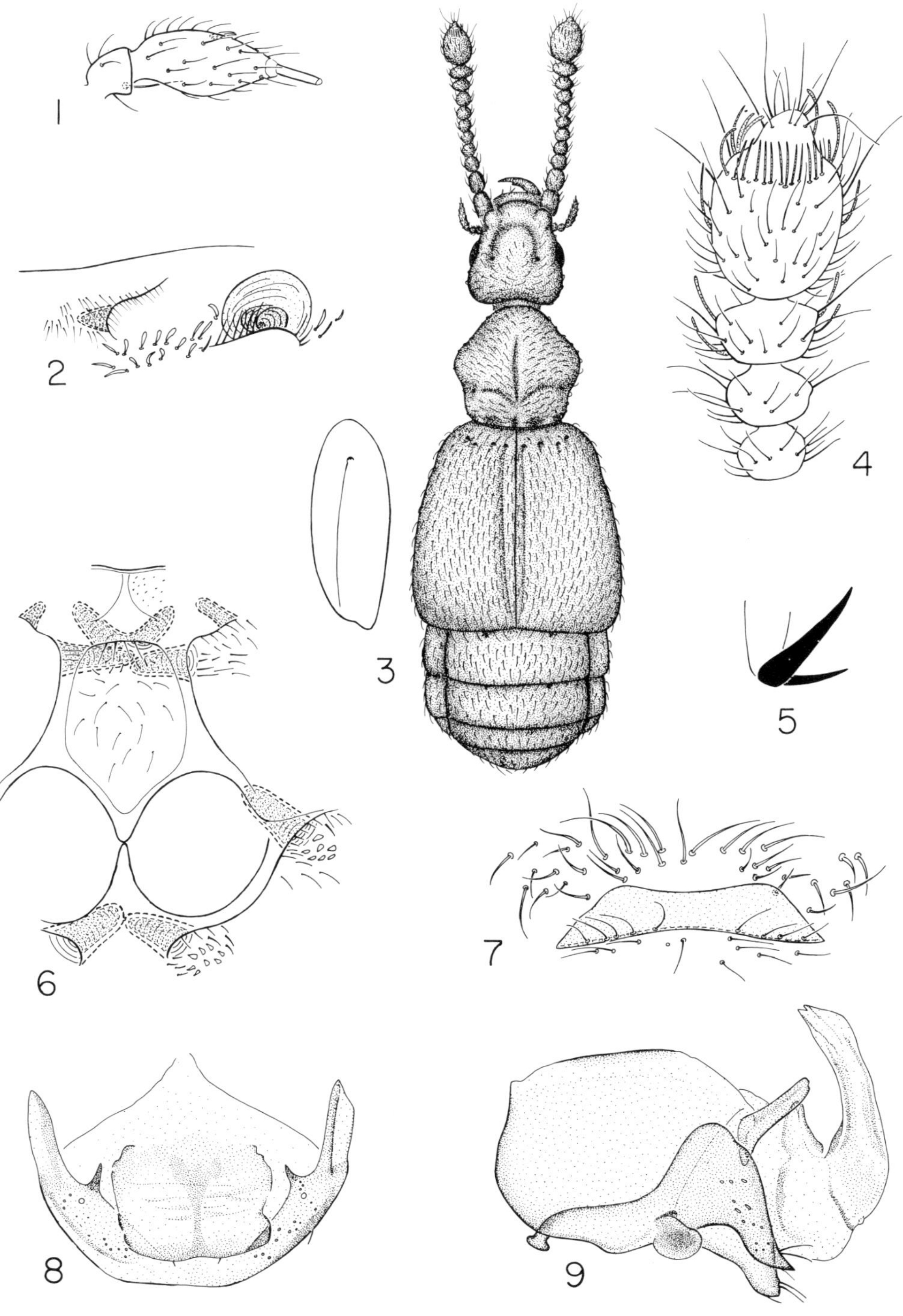

PLATE 29. *Oropodes orbiceps* Casey. Group B. Homotype ♂. Figs. 1–3, 7, 9, 10. Homotype ♀. Figs. 5, 6, 8. *Oropodes rumseyensis* Grigarick and Schuster. Paratype ♀. Fig. 4.

Fig. 1. Protarsal claws.
Fig. 2. Apex of mesotibia.
Fig. 3. Mesosternal area.
Fig. 4. Dorsal aspect.
Fig. 5. Antennal segments VIII to XI.
Fig. 6. Sternite VI.
Fig. 7. Sex-limited character of sternite III.
Fig. 8. Abdominal segments VIII and IX.
Fig. 9. Sternite VI.
Fig. 10. Genitalia, lateral aspect.

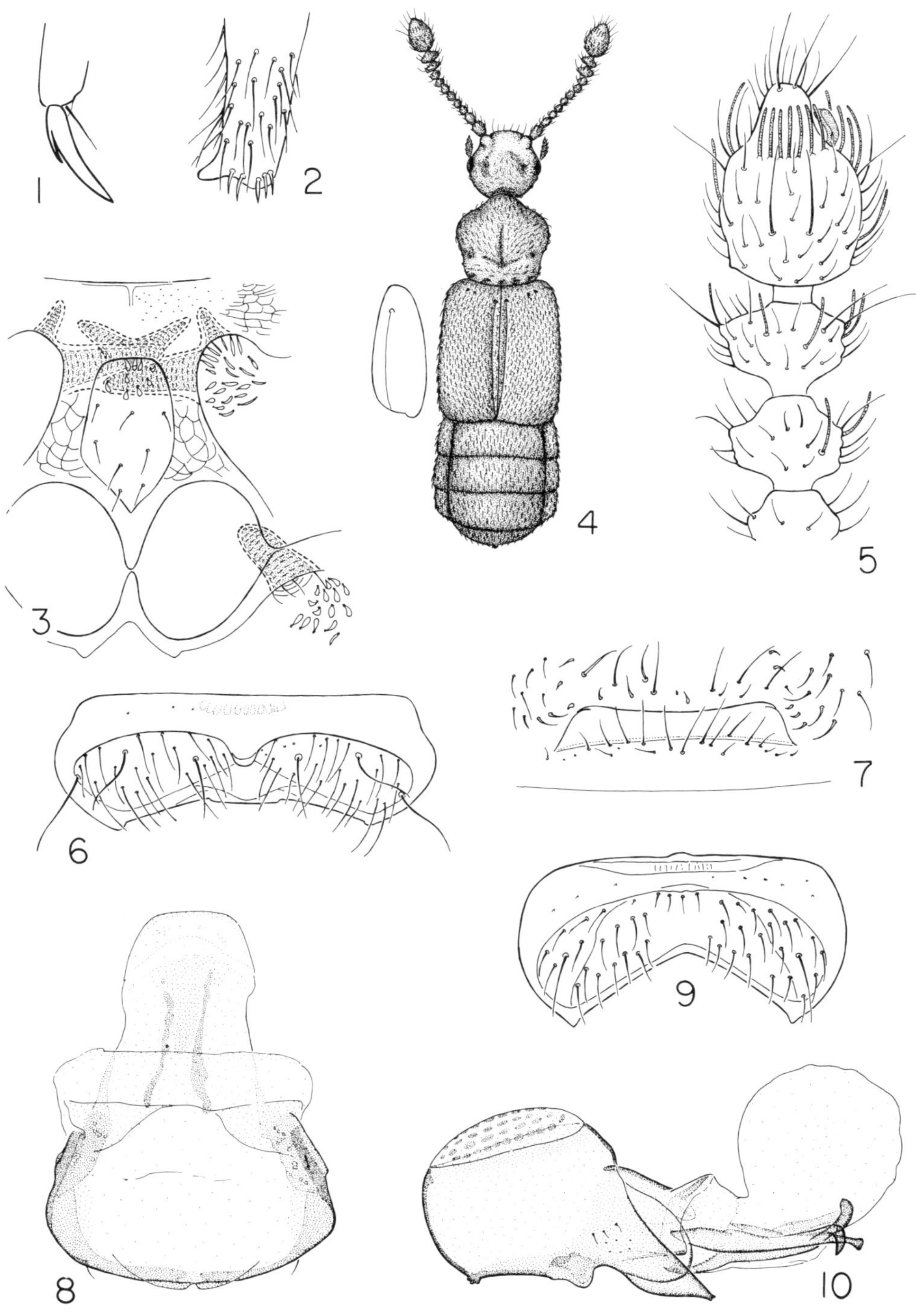

PLATE 30. *Trichonyx sulcicollis* (Reichenbach). Group B. ♀.

Fig. 1. Prosternal area.
Fig. 2. Dorsal aspect.
Fig. 3. Antennal segments VIII to XI.
Fig. 4. Mesosternal area.
Fig. 5. Mesotarsal claws.
Fig. 6. Abdominal segments VIII and IX.
Fig. 7. Metacoxae.
Fig. 8. Sternite II.

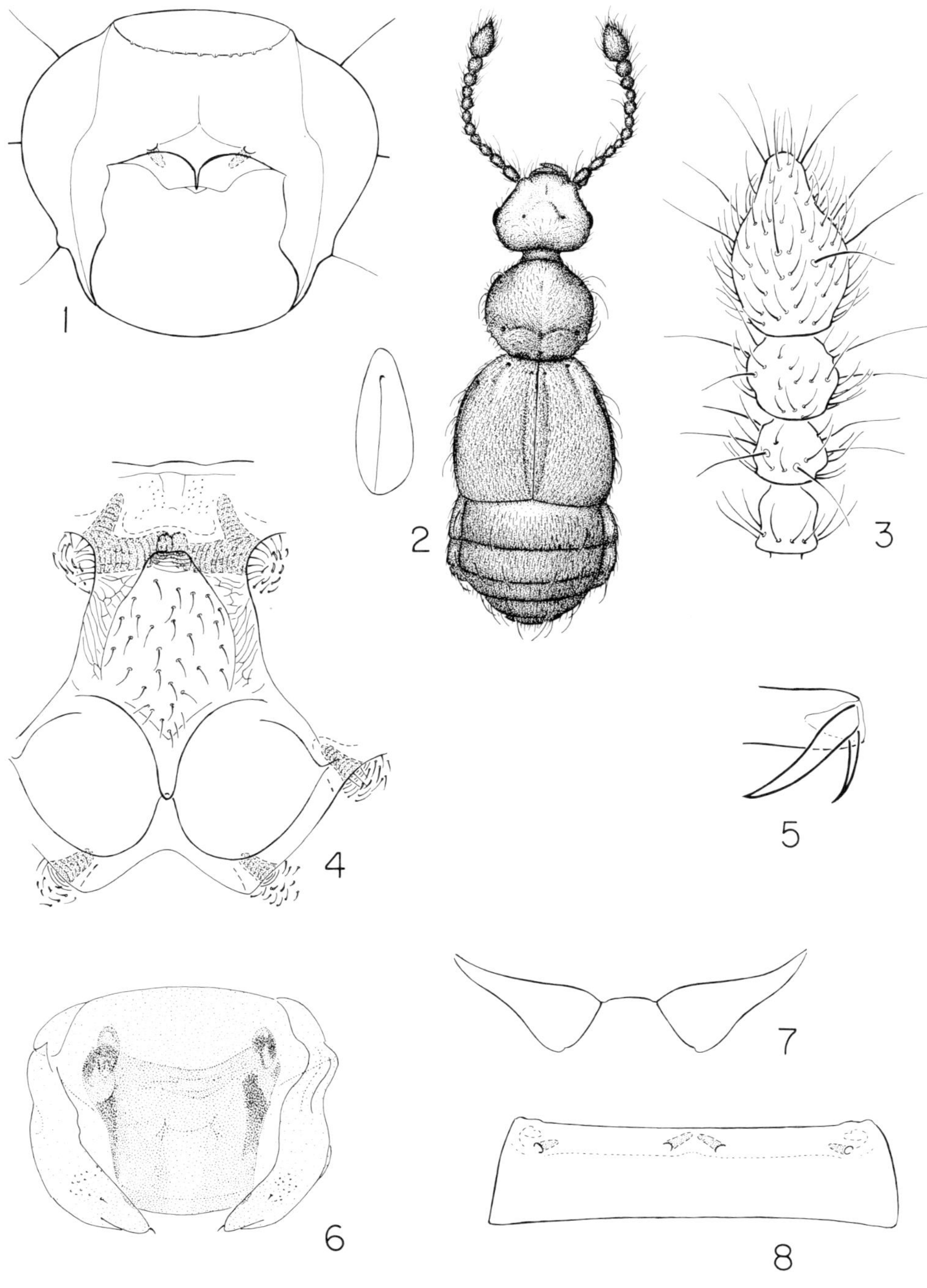

PLATE 31. *Abdiunguis fenderi* Park and Wagner. Group B. Homotypes ♂, ♀. Figs.
1-6, 8 ♂; fig. 7 ♀.

Fig. 1. Mesosternal area.
Fig. 2. Mesotarsal claws.
Fig. 3. Dorsal aspect.
Fig. 4. Antennal segments VIII to XI.
Fig. 5. Tergite III.
Fig. 6. Tergite IV.
Fig. 7. Abdominal segments VIII and IX.
Fig. 8. Genitalia, ventrolateral aspect.

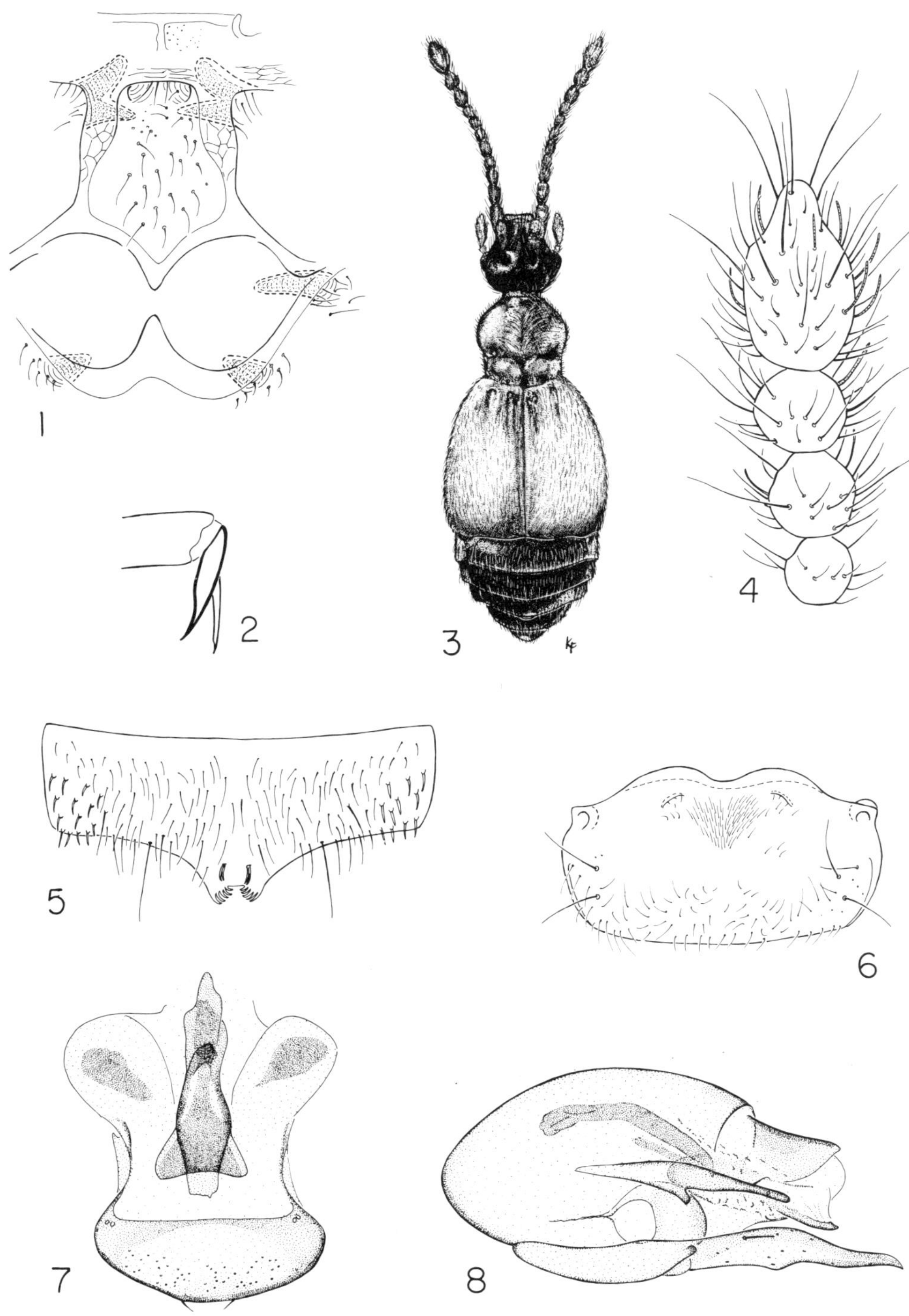

PLATE 32. *Euplecturga norstelcha* Grigarick and Schuster. Group B. Type ♂.

Fig. 1. Metatarsal claws.
Fig. 2. Sternite II.
Fig. 3. Dorsal aspect.
Fig. 4. Antennal segments IX to XI.
Fig. 5. Tergites I and II.
Fig. 6. Tergite VI.
Fig. 7. Mesosternal area.
Fig. 8. Genitalia, dorsal aspect.

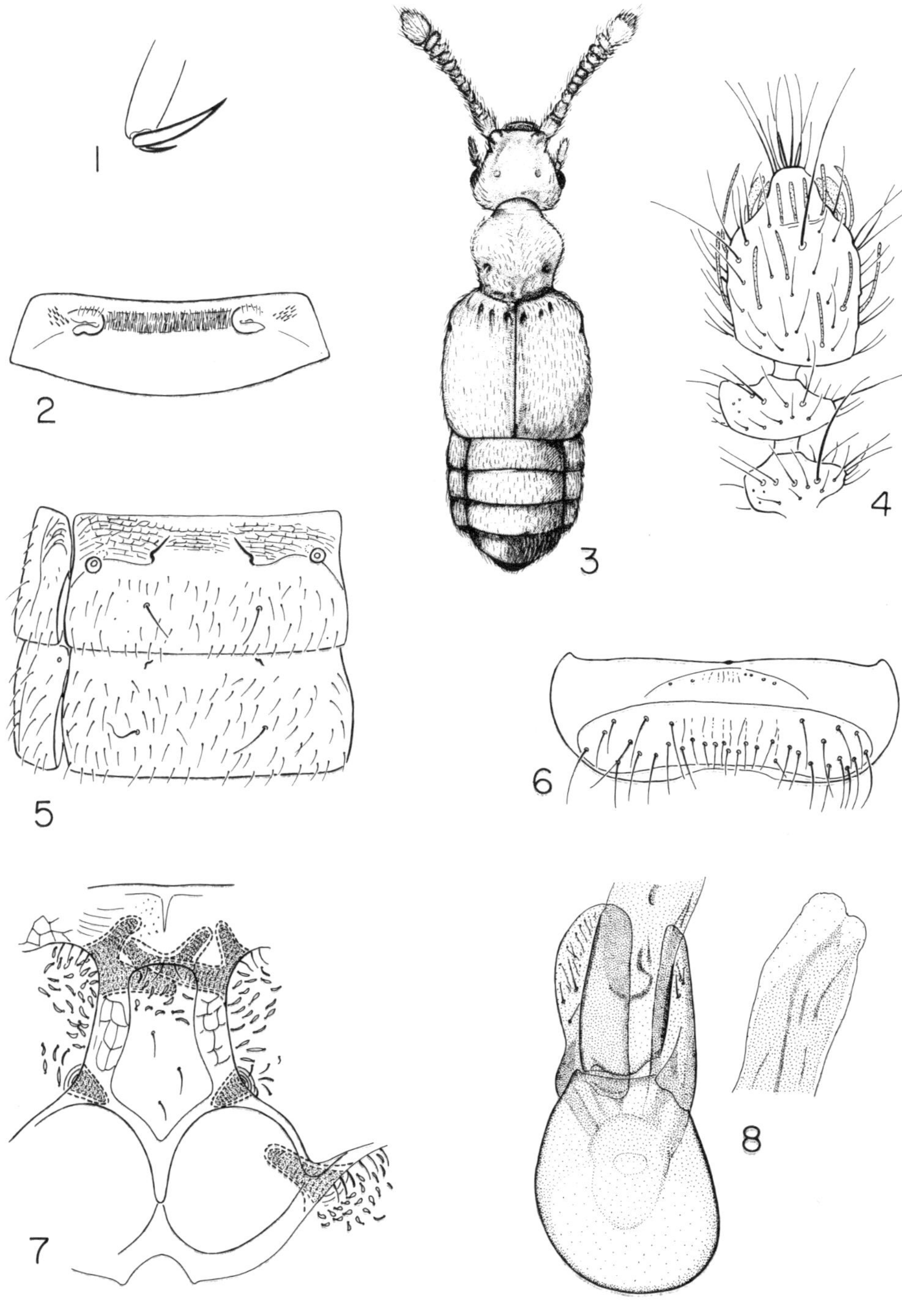

PLATE 33. *Tetrascapha dasycerca* Schuster and Marsh. Group B. Type ♂.

Fig. 1. Mesosternal area.
Fig. 2. Dorsal aspect.
Fig. 3. Antennal segments VII to XI.
Fig. 4. Sternite II.
Fig. 5. Mesotarsal claws.
Fig. 6. Sternite VI.
Fig. 7. Penial plate with one of two lateral plates.
Fig. 8. Genitalia, dorsal aspect.

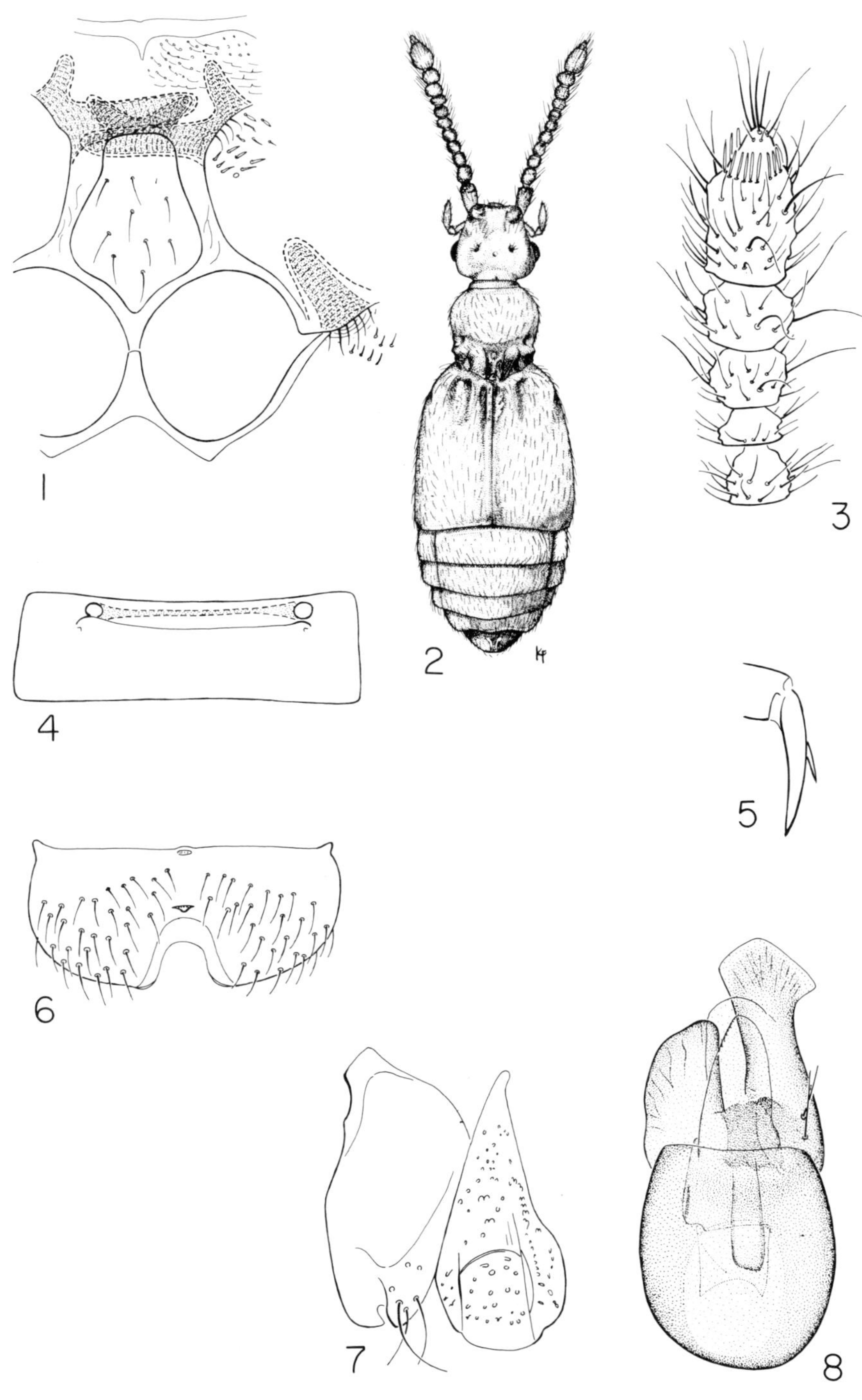

PLATE 34. *Mexiplectus emersoni* Park. Group B. Type ♂.

Fig. 1. Prosternum.
Fig. 2. Dorsal aspect.
Fig. 3. Maxillary palp IV.
Fig. 4. Antennal segments IX to XI.
Fig. 5. Mesosternal area.
Fig. 6. Metatarsal claw.
Fig. 7. Pro-tergite I.
Fig. 8. Genitalia, dorsal aspect.

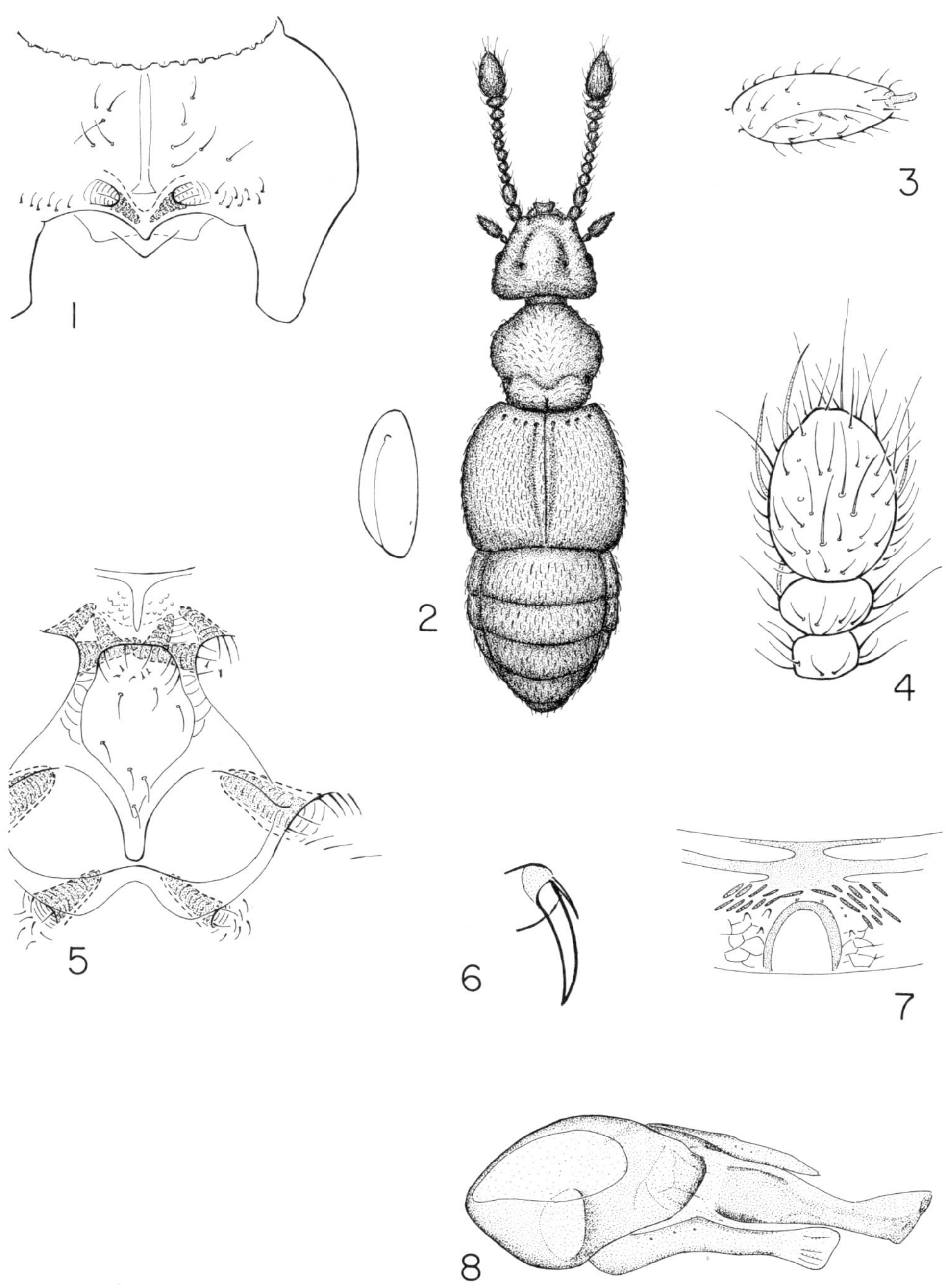

PLATE 35. *Verabarolus subdendrus* Park. Group B. Type ♀.

Fig. 1. Prosternum.
Fig. 2. Dorsal aspect.
Fig. 3. Antennal segments IX to XI.
Fig. 4. Mesosternal area.
Fig. 5. Pro-tergite I.

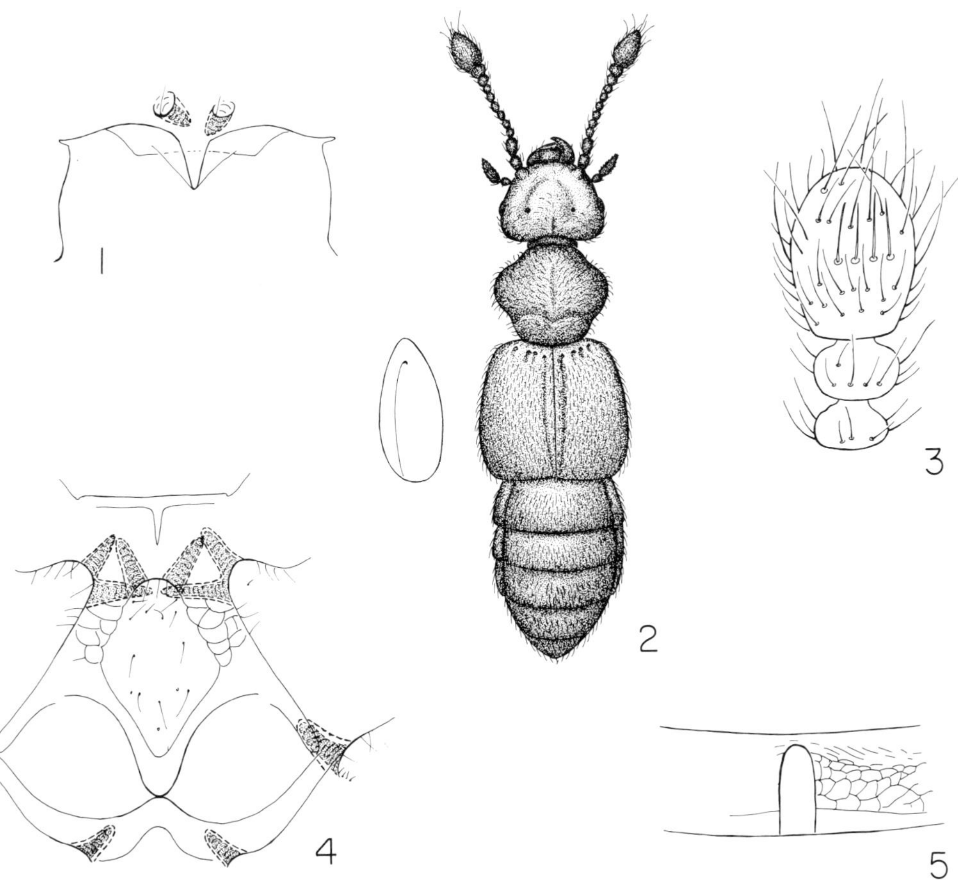

PLATE 36. *Thesiectus probus* Park. Group B. Paratype ♀. Figs. 1–5. Type ♂. Fig. 6.

Fig. 1. Prosternum.
Fig. 2. Mesosternal area.
Fig. 3. Dorsal aspect.
Fig. 4. Antennal segments IX to XI.
Fig. 5. Mesotarsal claw.
Fig. 6. Genitalia, dorsal aspect.

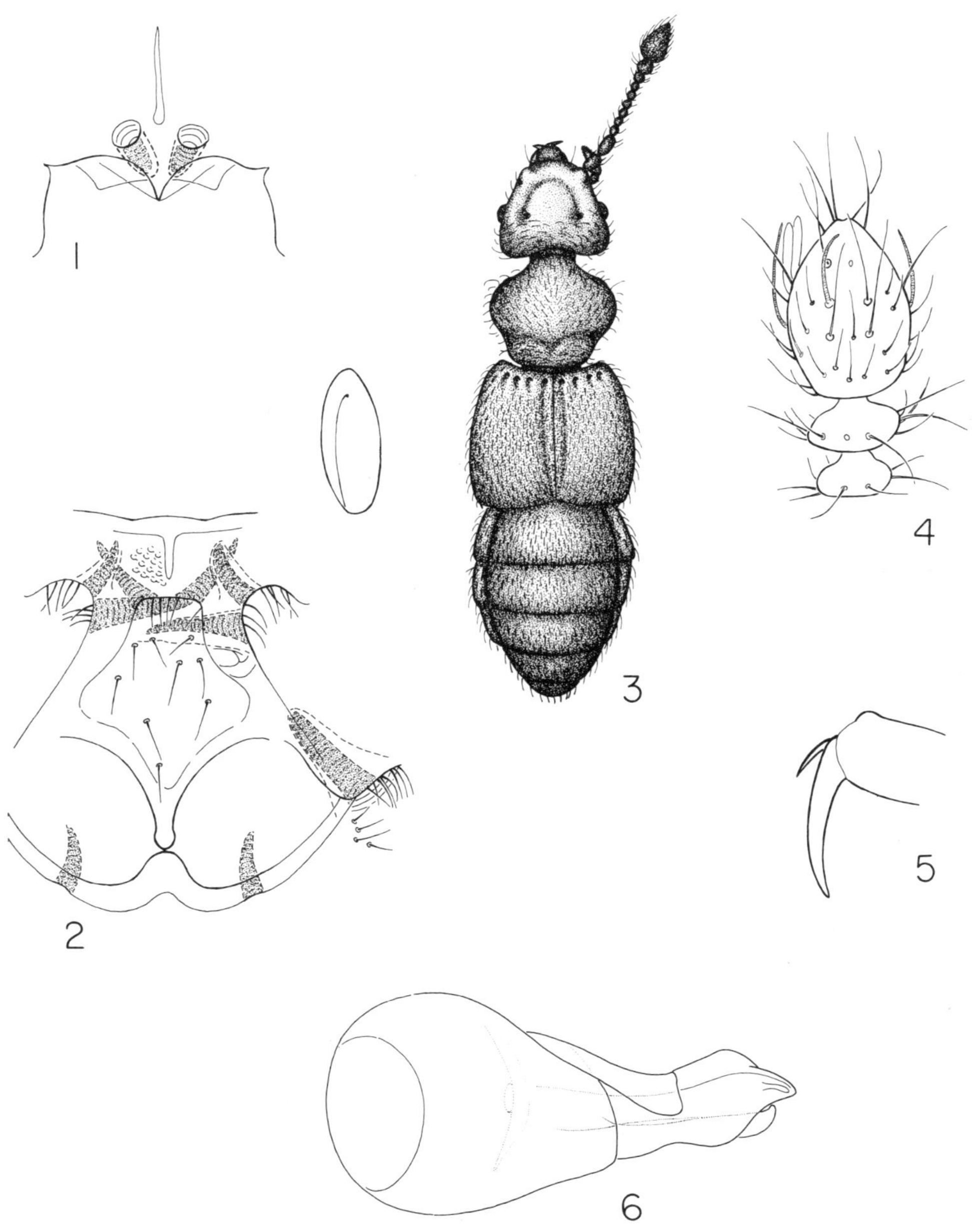

PLATE 37. *Panaramecia williamsi* Park. Group B. Type ♂.

Fig. 1. Prosternal area.
Fig. 2. Dorsal aspect.
Fig. 3. Antennal segments VIII to XI.
Fig. 4. Mesosternal area.
Fig. 5. Labrum, ventral.
Fig. 6. Metatarsal claw.
Fig. 7. Genitalia, dorsal aspect.

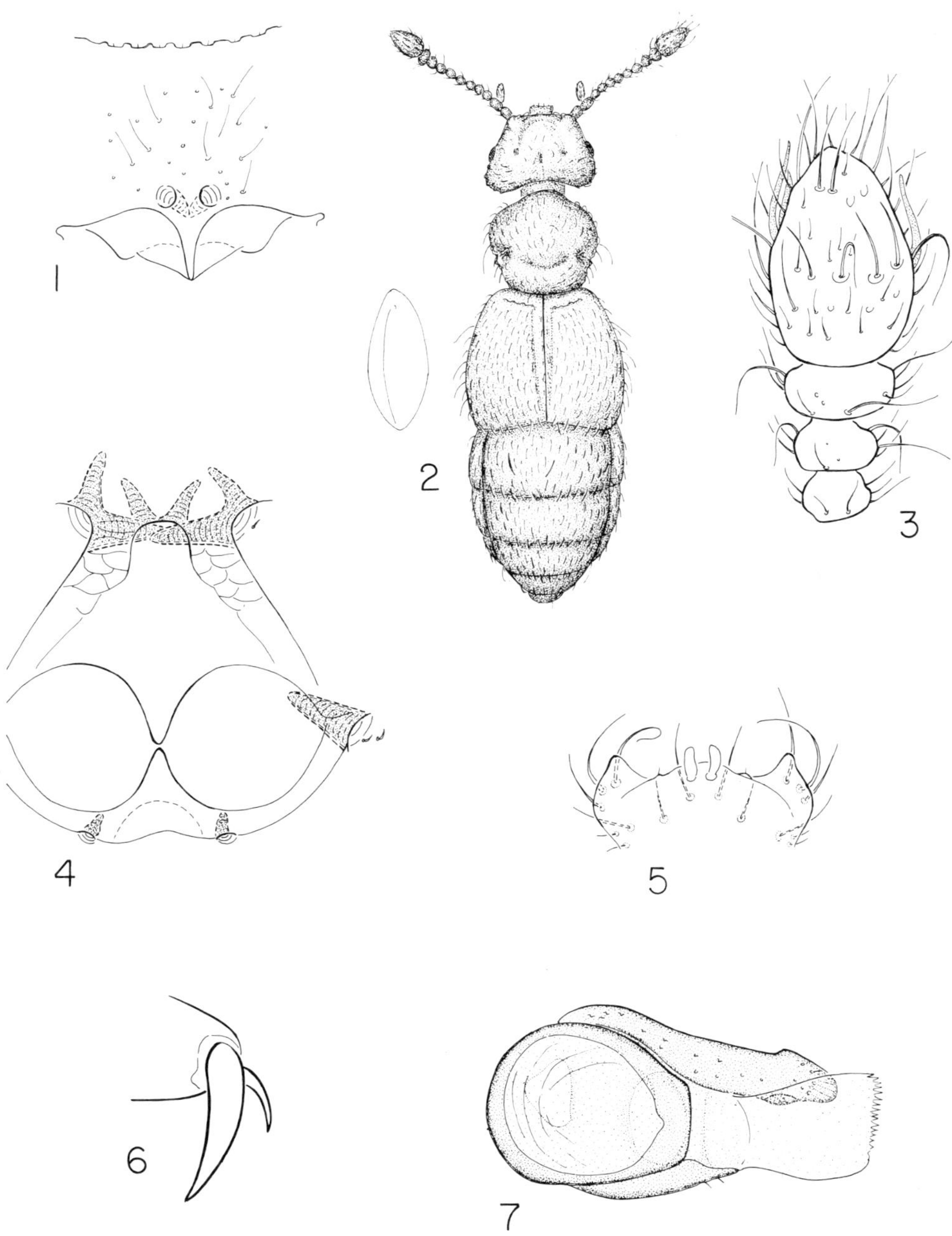

PLATE 38. *Mexigaster boneti* Park. Group C. Type ♂.

Fig. 1. Prosternal area.
Fig. 2. Dorsal aspect.
Fig. 3. Antennal segments IX to XI.
Fig. 4. Metatarsal claws.
Fig. 5. Mesosternal area.
Fig. 6. Setae on median of tergite II.
Fig. 7. Sternite II.
Fig. 8. Profemur.
Fig. 9. Genitalia, lateral aspect.

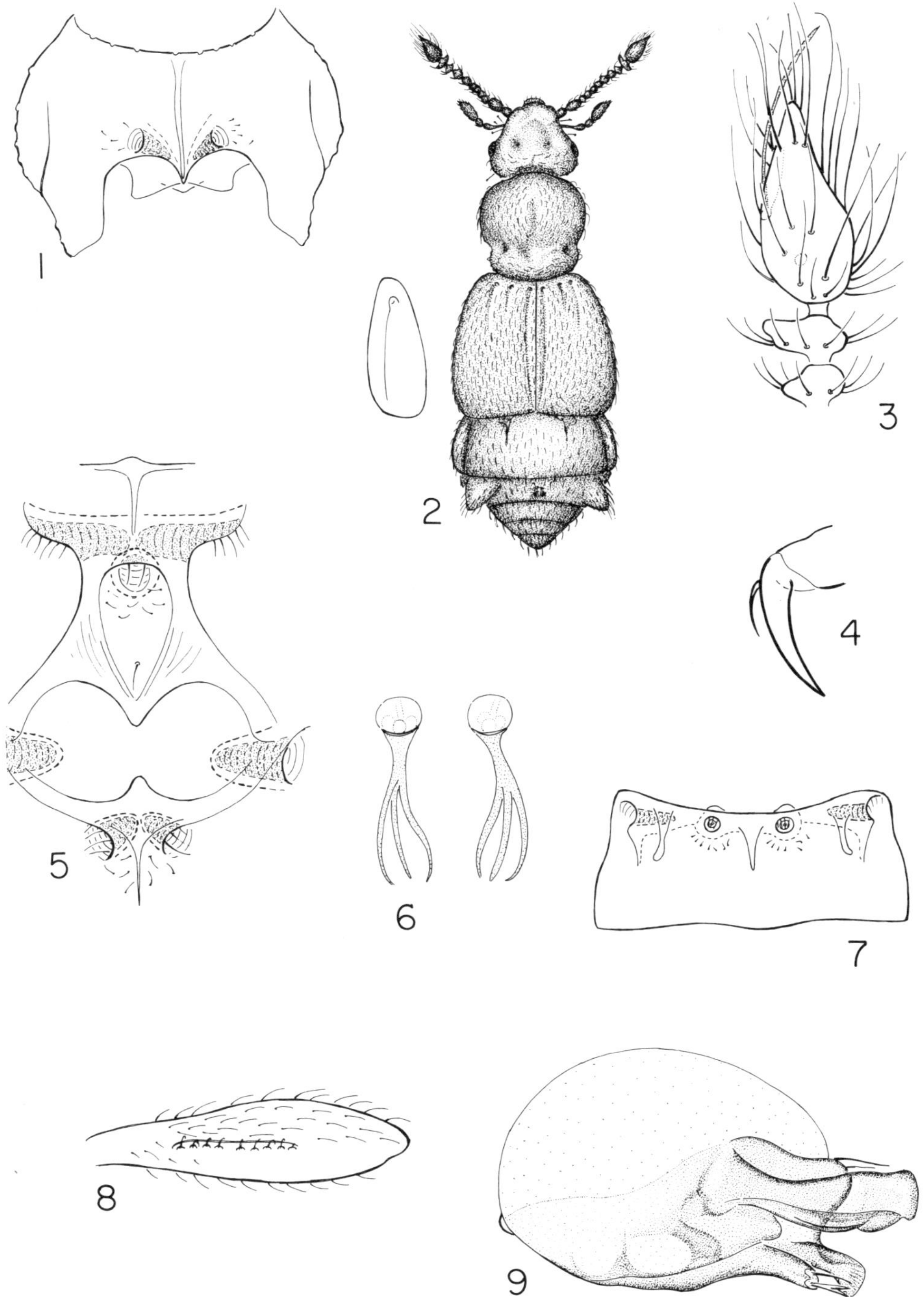

PLATE 39. *Leptoplectus pertenuis* (Casey). Group C. Homotype ♂.

Fig. 1. Ventral aspect of head.
Fig. 2. Dorsal aspect.
Fig. 3. Labrum.
Fig. 4. Prosternum.
Fig. 5. Apex of abdomen.
Fig. 6. Antennal segments VIII to XI.
Fig. 7. Mesosternal area.
Fig. 8. Genitalia, dorsal aspect.

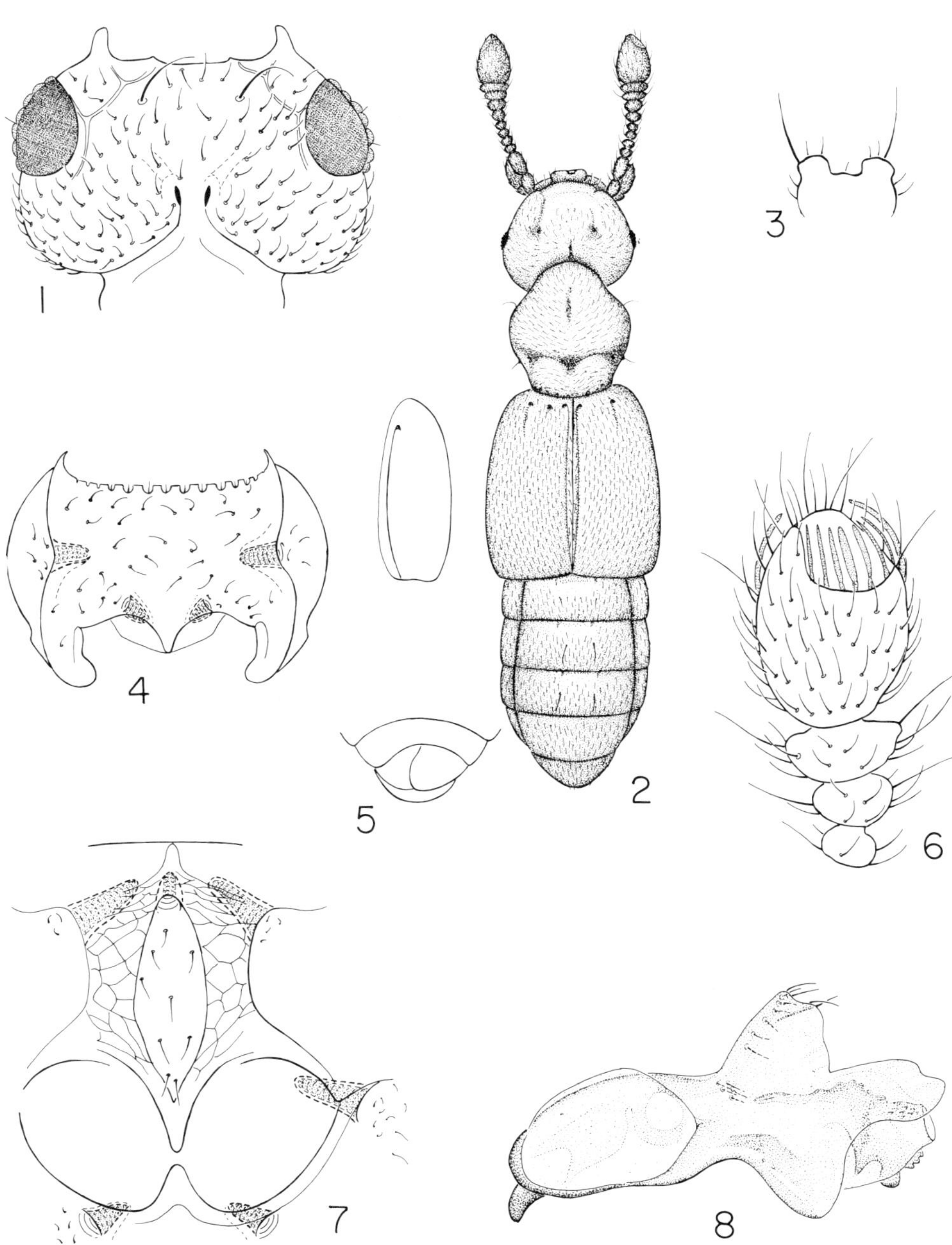

PLATE 40. *Biblomimus vertexalis* Park. Group C. Type ♂.

Fig. 1. Head, dorsal aspect.
Fig. 2. Head, ventral aspect.
Fig. 3. Dorsal aspect.
Fig. 4. Antennal segments IX to XI.
Fig. 5. Metatarsal claw.
Fig. 6. Mesosternal area.
Fig. 7. Profemur.
Fig. 8. Prosternal area.
Fig. 9. Genitalia, lateral aspect.

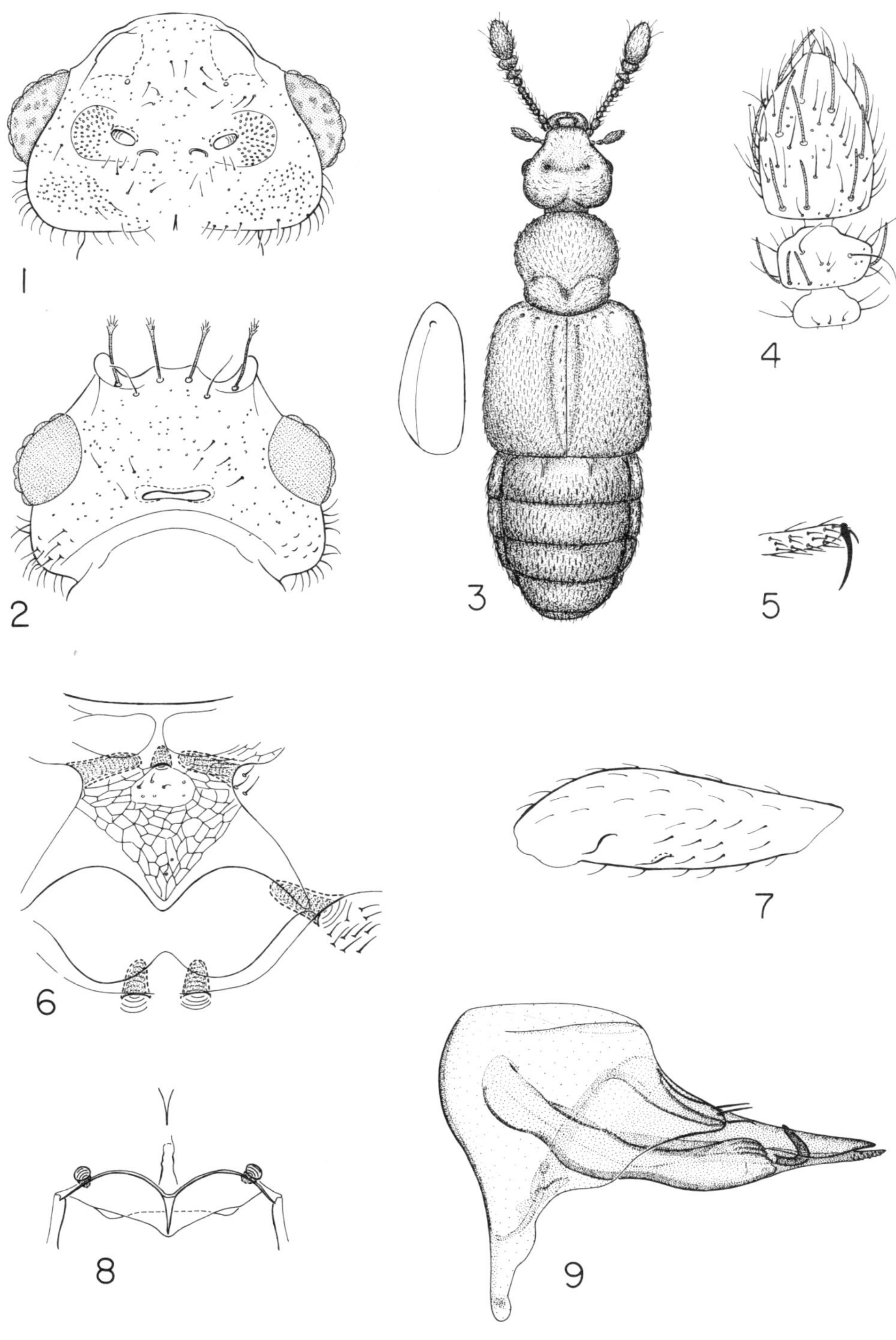

PLATE 41. *Liniolis crinitus* (Brendel). Group C. Lectotype ♂.

Fig. 1. Prosternum.
Fig. 2. Dorsal aspect.
Fig. 3. Profemur.
Fig. 4. Mesosternal area.
Fig. 5. Metatrochanter.
Fig. 6. Tergites II to IV.
Fig. 7. Genitalia, lateral aspect.

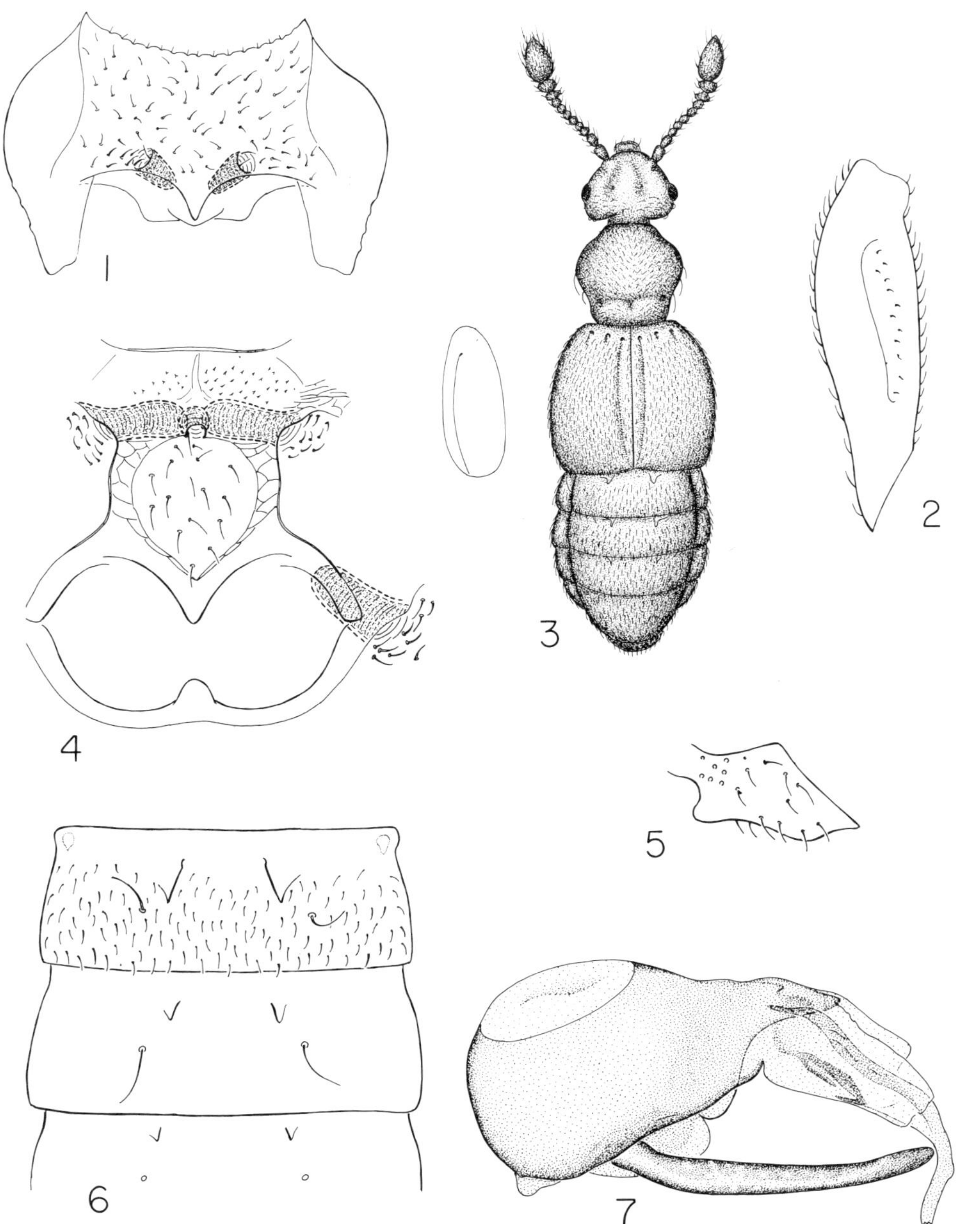

PLATE 42. *Tomoplectus mexicanus* Park. Group C. Type ♀. Fig. 2. Paratype ♀. Figs. 1, 3–8.

Fig. 1. Maxillary palp.
Fig. 2. Dorsal aspect.
Fig. 3. Antennal segments VIII to XI.
Fig. 4. Mesosternal area.
Fig. 5. Mesotarsal claws.
Fig. 6. Profemur.
Fig. 7. Abdominal segments VIII and IX.
Fig. 8. Tergite I.

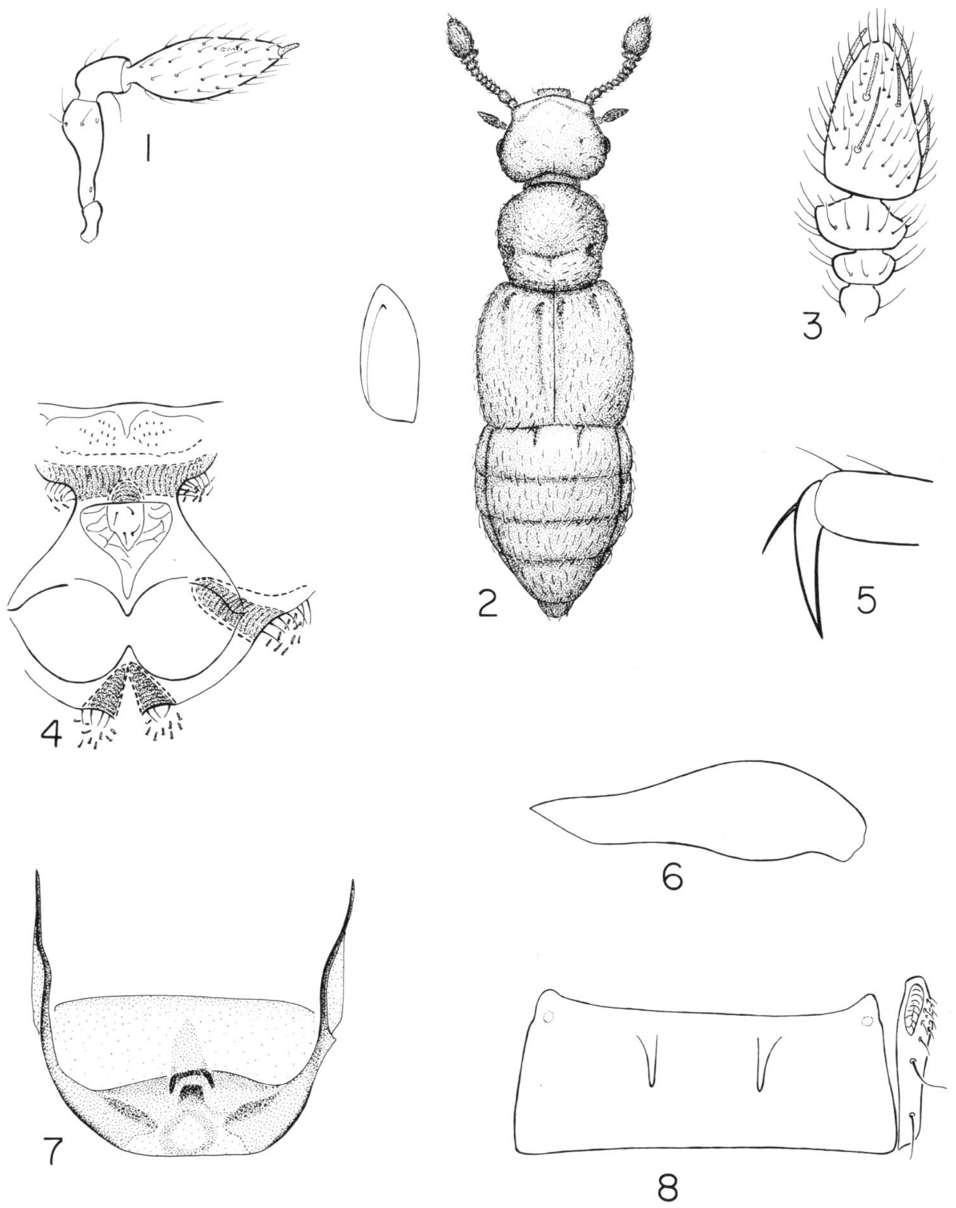

PLATE 43. *Cupila clavicornis* (Mäklin). Group C. Figs. 1, 4, 5, 7 ♂. *Cupila multifossa* Grigarick and Schuster. Figs. 2-3 ♀. *Cupila excavata* Park and Wagner. Fig. 6 ♀.

Fig. 1. Prosternal area.
Fig. 2. Dorsal aspect.
Fig. 3. Antennal segments VIII to XI.
Fig. 4. Mesosternal area.
Fig. 5. Mesotarsal claws.
Fig. 6. Abdominal segments VIII and IX.
Fig. 7. Genitalia, dorsal aspect.

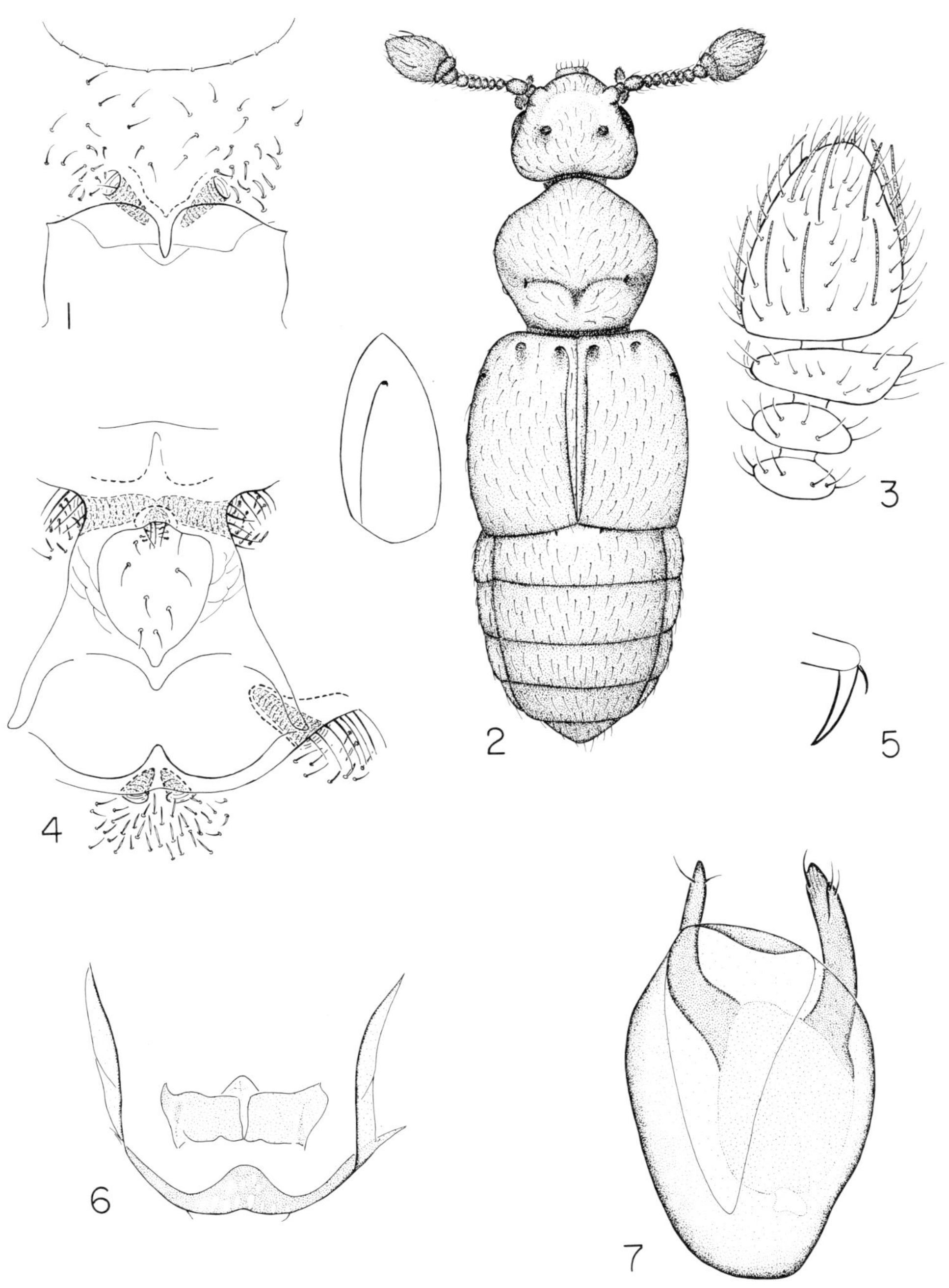

PLATE 44. *Pilactium summersi* Grigarick and Schuster. Group C. Type ♂. Figs. 1-7, 9. Fig. 8 ♀.

Fig. 1. Prosternal area.
Fig. 2. Mesotarsal claws.
Fig. 3. Antennal segments IX to XI.
Fig. 4. Mesosternal area.
Fig. 5. Tergite I.
Fig. 6. Profemur.
Fig. 7. Sternite VI.
Fig. 8. Abdominal segments VIII and IX.
Fig. 9. Genitalia, lateral aspect.

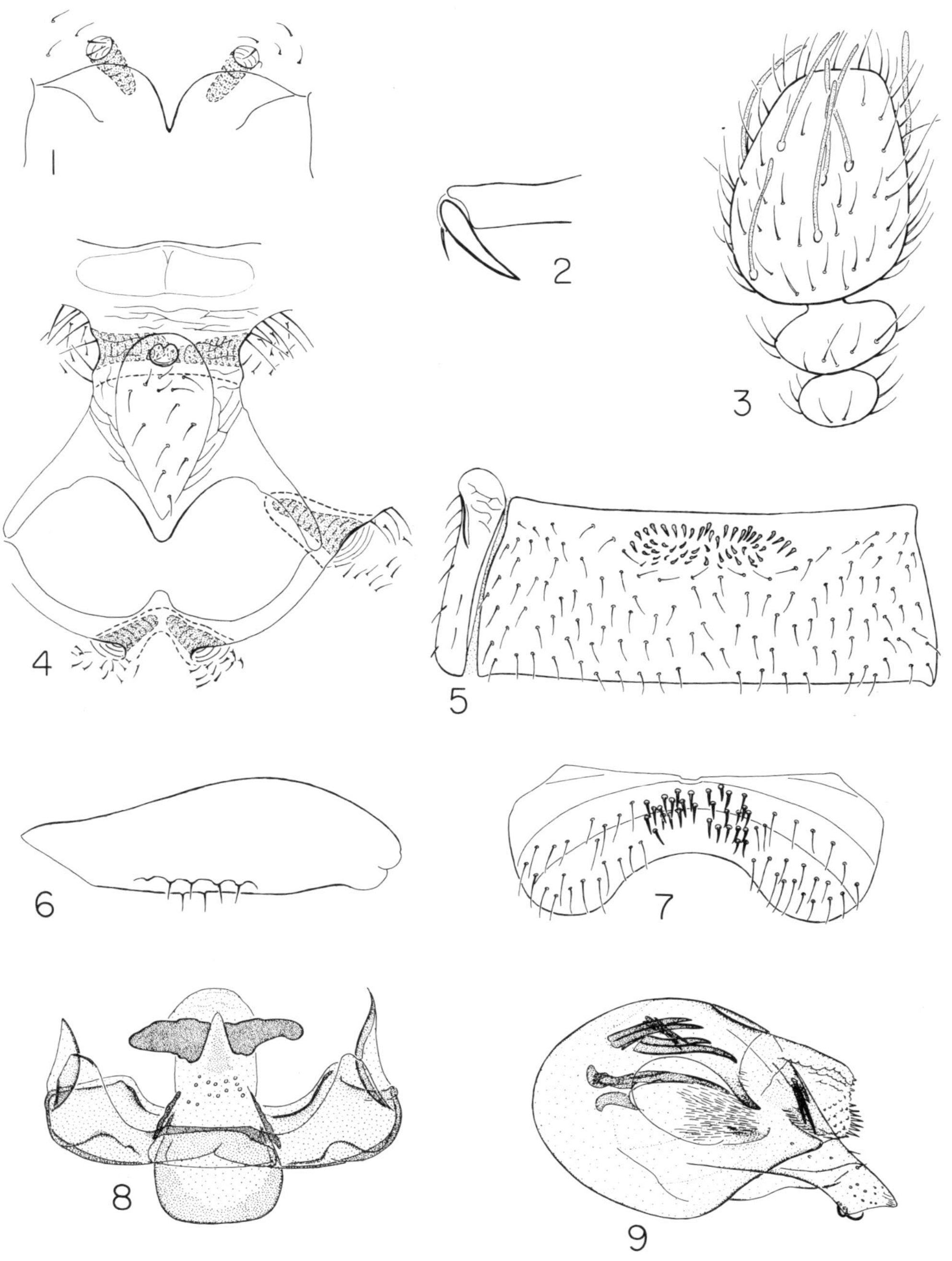

PLATE 45. *Pygmactium steevesi* Schuster and Grigarick. Group C. Type ♂. Figs. 1–7, 9. *Pygmactium mollyae* (Park). Fig. 8 ♀.

Fig. 1. Prosternum.
Fig. 2. Dorsal aspect.
Fig. 3. Antennal segments IX to XI.
Fig. 4. Prosternal area.
Fig. 5. Mesotarsal claws.
Fig. 6. Profemur.
Fig. 7. Tergite I.
Fig. 8. Abdominal segments VIII and IX.
Fig. 9. Genitalia, dorsal aspect.

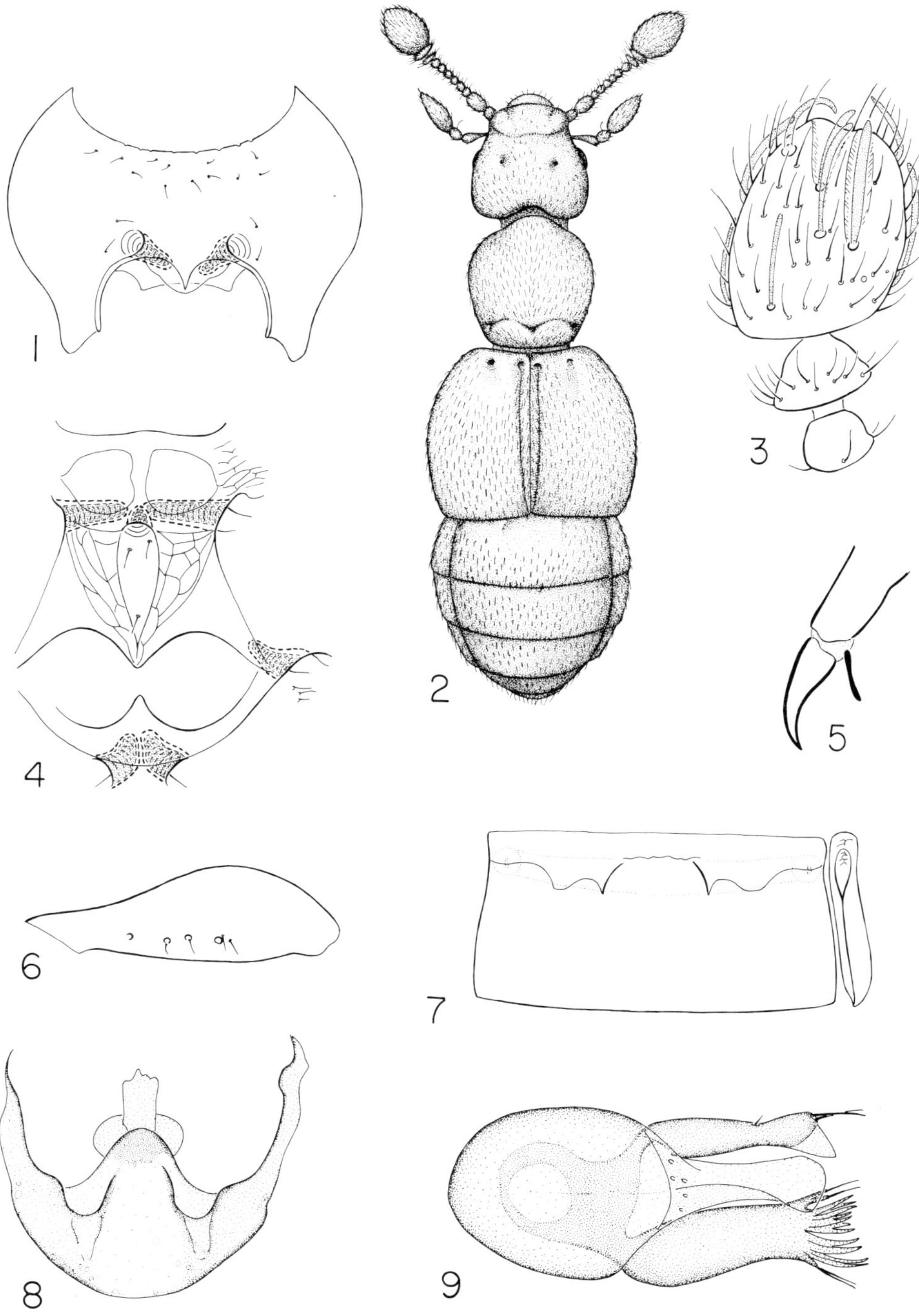

PLATE 46. *Actium californicum* (LeConte). Group C. Figs. 1–5, 7 ♂; fig. 6 ♀.

Fig. 1. Mesosternal area.
Fig. 2. Mesostarsal claws.
Fig. 3. Dorsal aspect.
Fig. 4. Antennal segments VIII to XI.
Fig. 5. Tergite I.
Fig. 6. Abdominal segments VIII and IX.
Fig. 7. Genitalia, lateral aspect.

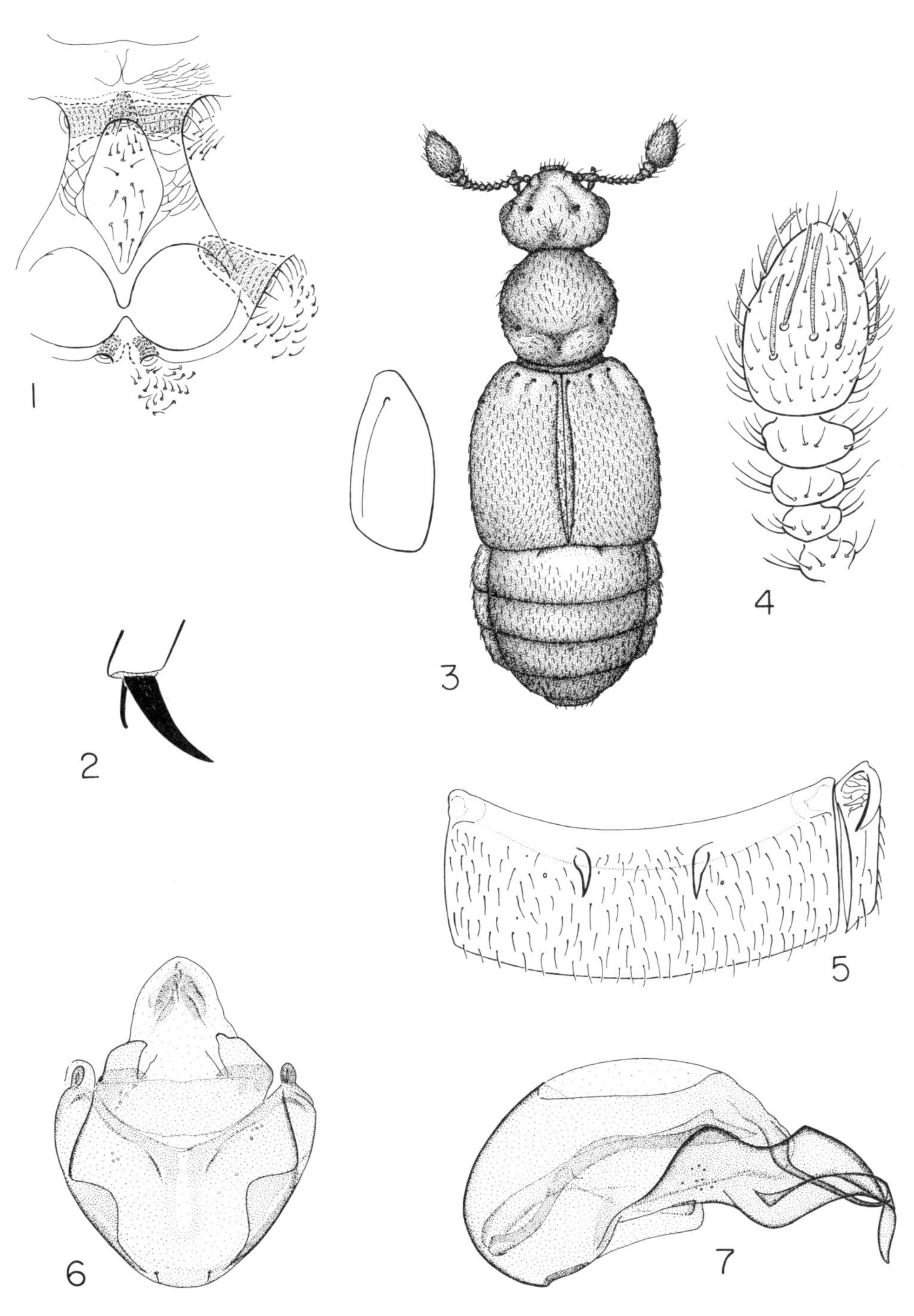

PLATE 47. *Euplectamecia hidalgensis* Park. Group C. Type ♂.

Fig. 1. Prosternal area.
Fig. 2. Dorsal aspect.
Fig. 3. Antennal segments IX to XI.
Fig. 4. Mesosternal area.
Fig. 5. Tergite I, setae omitted.
Fig. 6. Sternites II and III.
Fig. 7. Protarsal segments and claw.
Fig. 8. Genitalia, dorsal aspect.

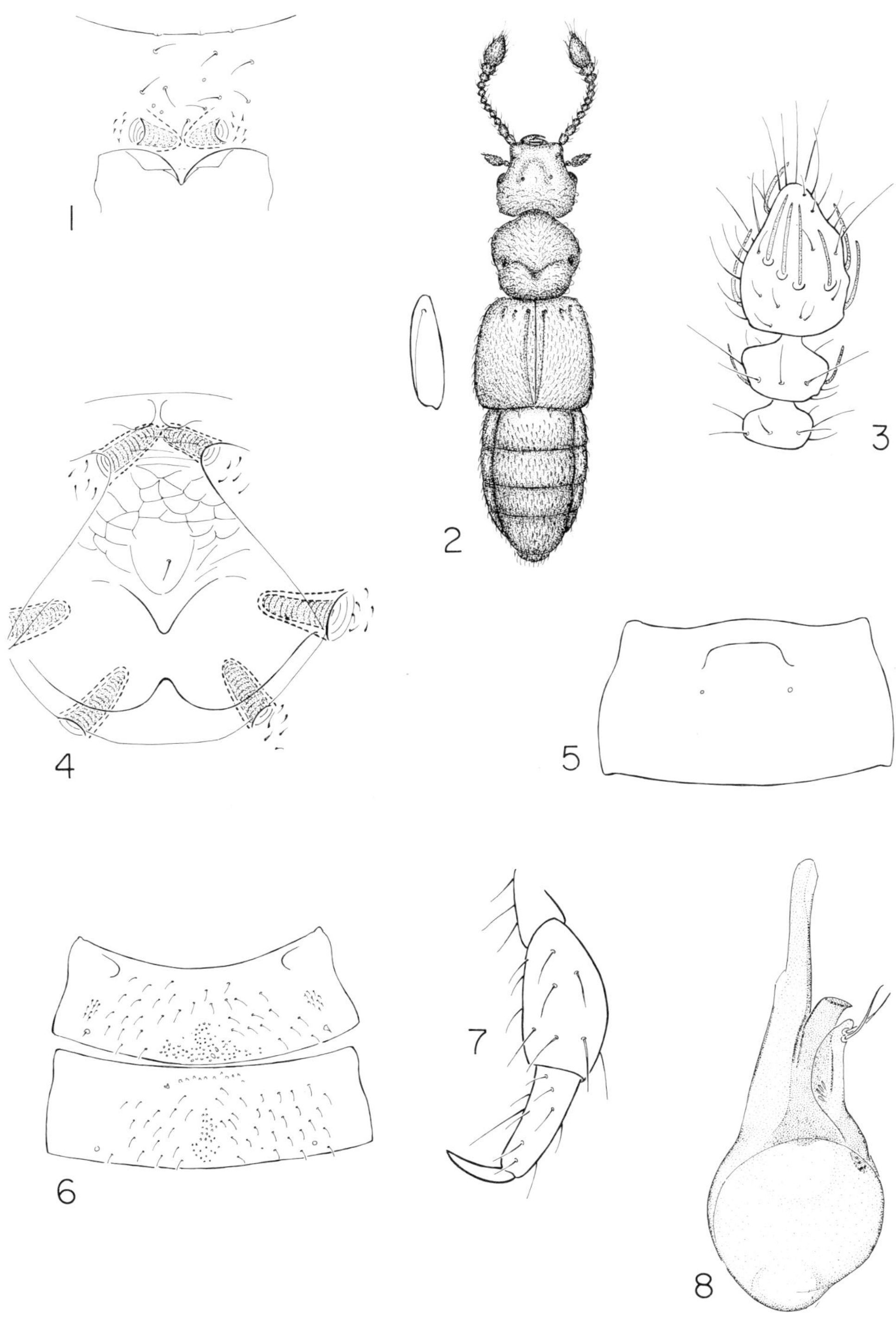

PLATE 48. *Thesiastes* sp. Group C. ♂.

Fig. 1. Mesosternal area.
Fig. 2. Dorsal aspect.
Fig. 3. Antennal segments IX to XI.
Fig. 4. Metatarsal claw.
Fig. 5. Tergites I to III.
Fig. 6. Profemur.
Fig. 7. Genitalia, lateral aspect.

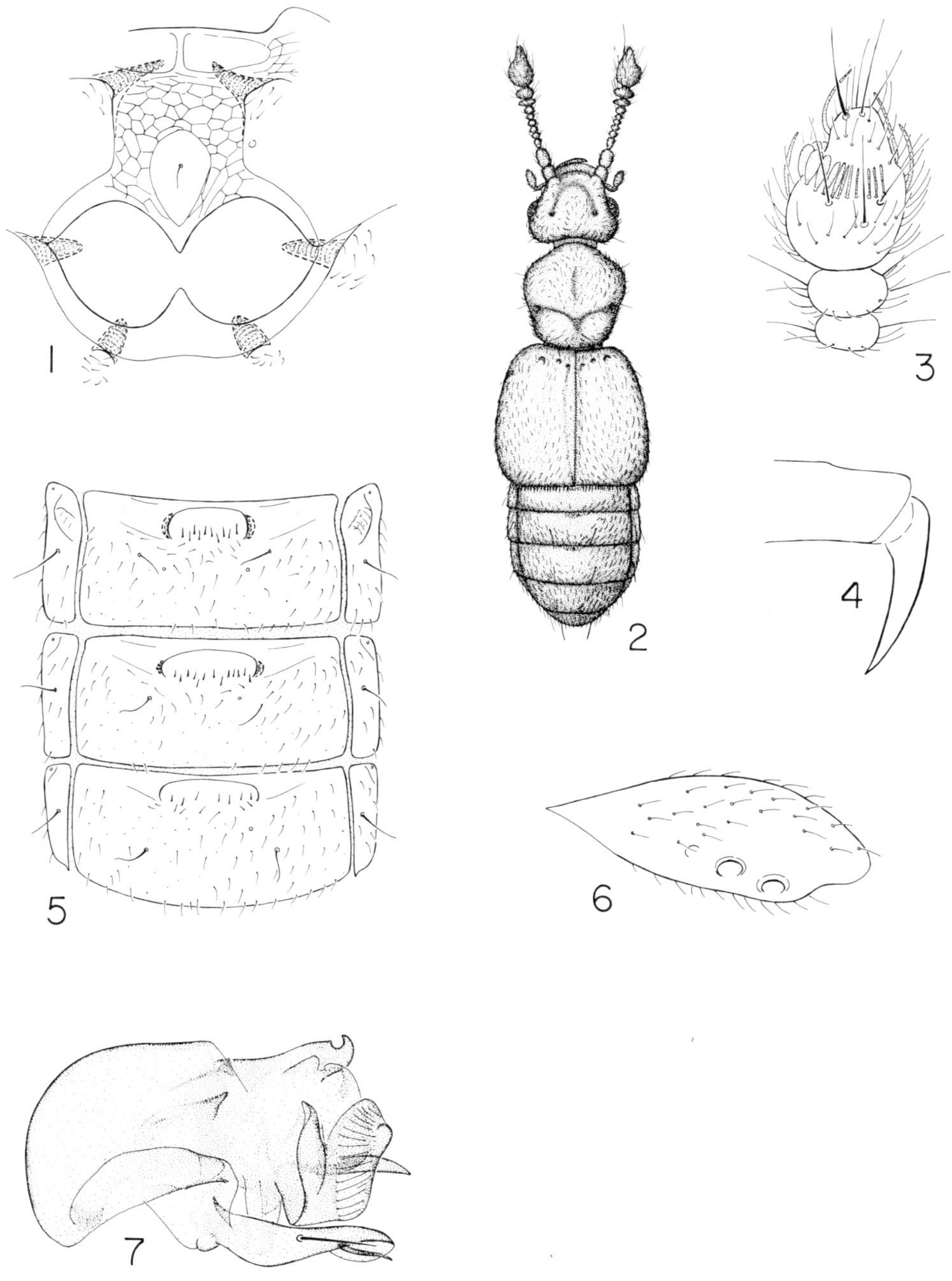

PLATE 49. *Bibloplectus sobrinus* Casey. Group C. Paratype ♂, ♀. Figs. 1-6 ♂; 7 ♀.

Fig. 1. Mesosternal area.
Fig. 2. Dorsal aspect.
Fig. 3. Antennal segments VI to XI.
Fig. 4. Tergite I and anterior margin of tergite II.
Fig. 5. Tarsal claw.
Fig. 6. Genitalia, lateral aspect.
Fig. 7. Abdominal segments VIII and IX.

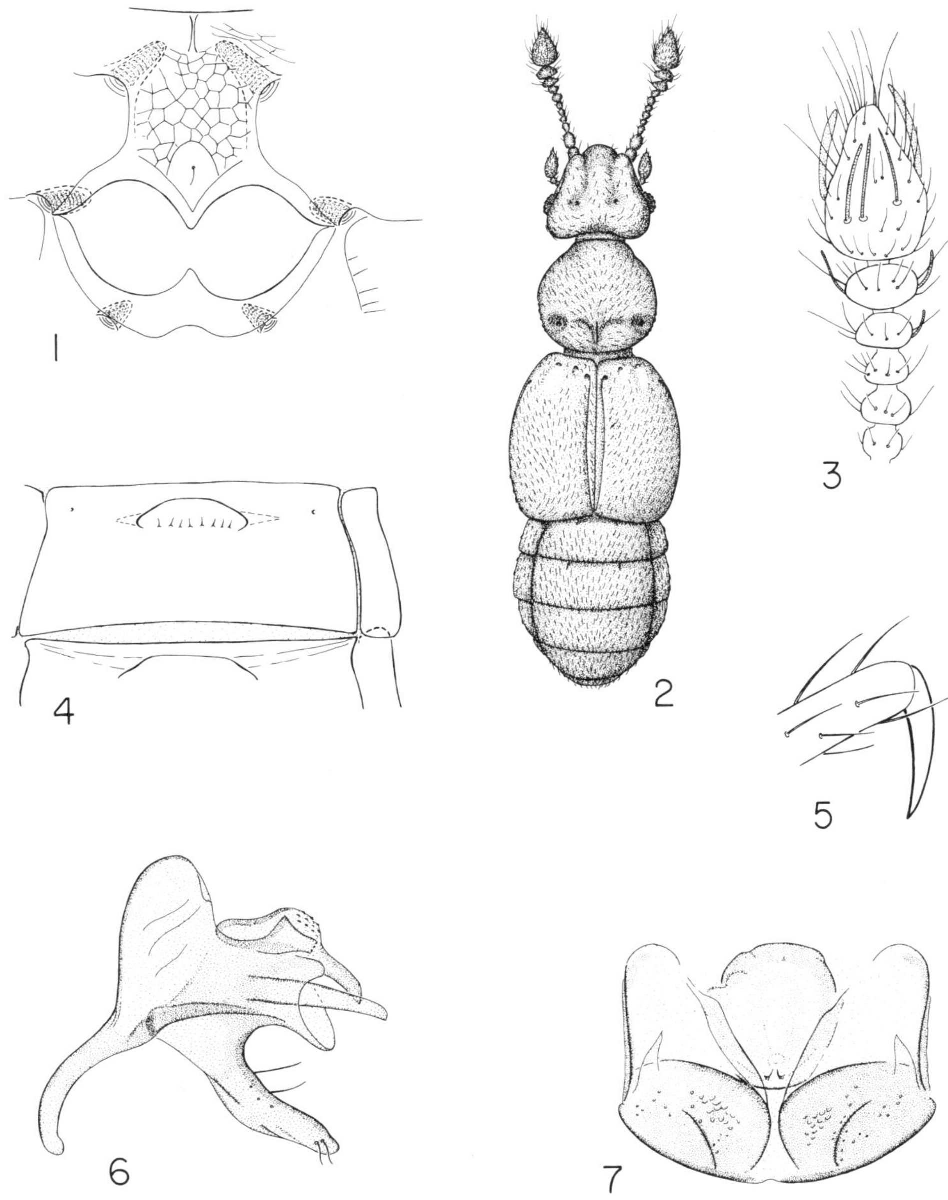

PLATE 50. *Saxet decoris* (Casey). Group C. Type ♀. Figs. 1-2, 5-6, 8. Homotype ♂.
Figs. 3-4, 7, 9.

Fig. 1. Mesosternal area.
Fig. 2. Dorsal aspect.
Fig. 3. Antennal segments VIII to XI.
Fig. 4. Tergites I to III.
Fig. 5. Mesotarsal claws.
Fig. 6. Femur and mesotrochanter.
Fig. 7. Metatrochanter.
Fig. 8. Abdominal segments VIII and IX.
Fig. 9. Genitalia, lateral aspect.

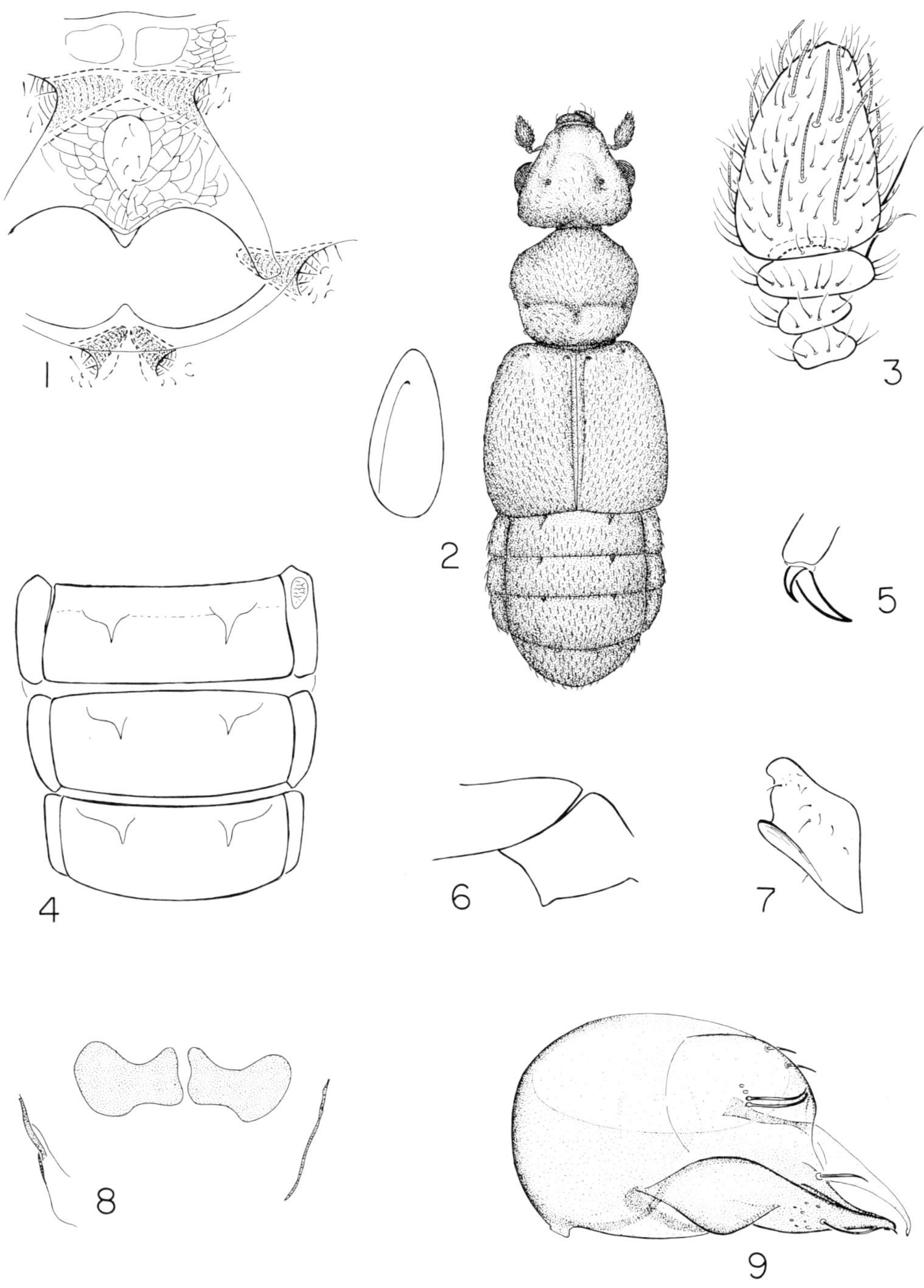

PLATE 51. *Dalmomelba cazieri* Park. Group C. Type ♂.

Fig. 1. Prosternum.
Fig. 2. Dorsal aspect.
Fig. 3. Antennal segments IX to XI.
Fig. 4. Mesosternal area.
Fig. 5. Mesotarsal claw.
Fig. 6. Modifications on tergites III and IV.
Fig. 7. Genitalia, lateral aspect.

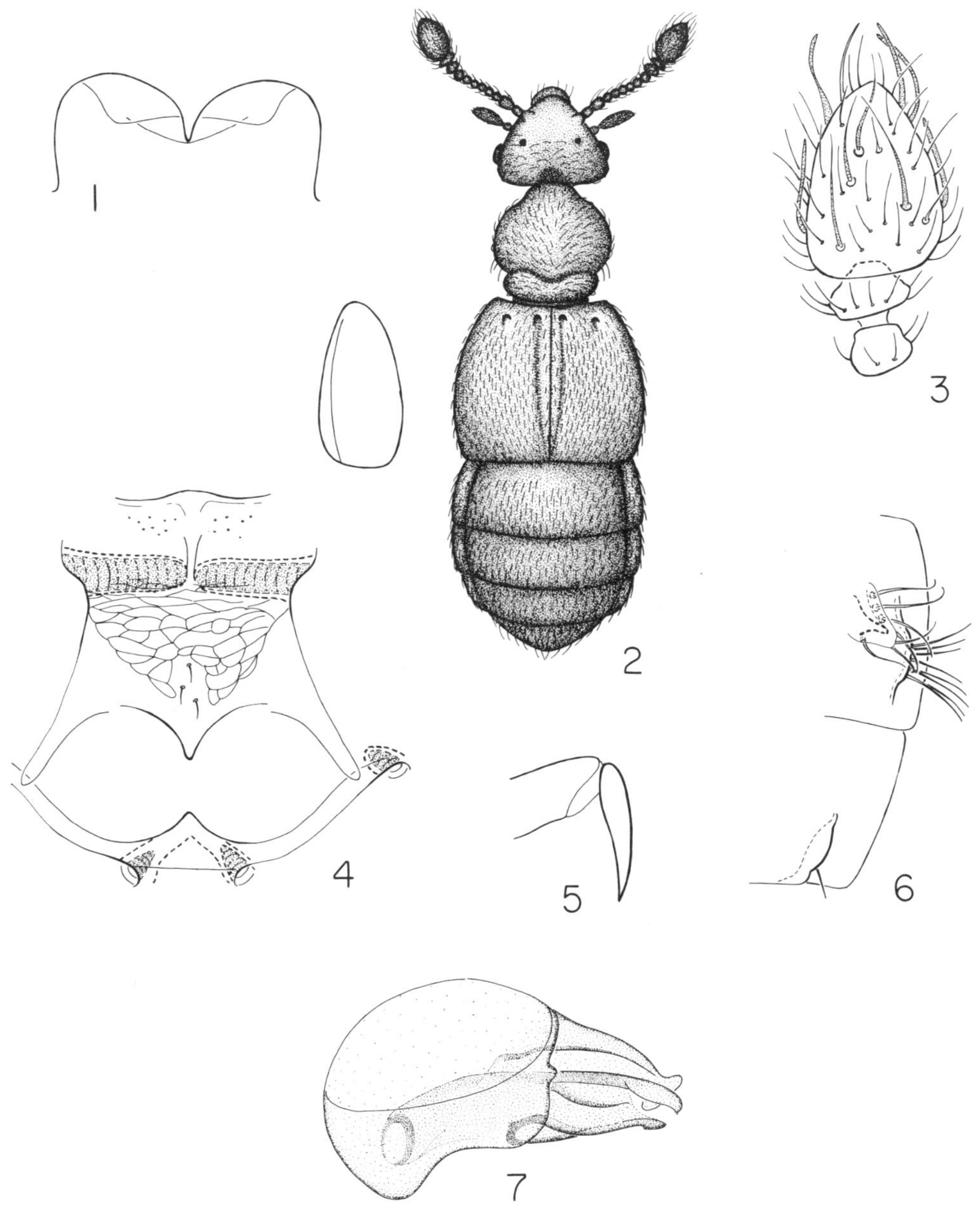

PLATE 52. *Trimiopsis furcalis* Park. Group C. Type ♂.

Fig. 1. Prosternal area.
Fig. 2. Dorsal aspect.
Fig. 3. Antennal segments IX to XI.
Fig. 4. Mesosternal area.
Fig. 5. Metatarsal claws.
Fig. 6. Mesotrochanter and femur.
Fig. 7. Genitalia, lateral aspect.

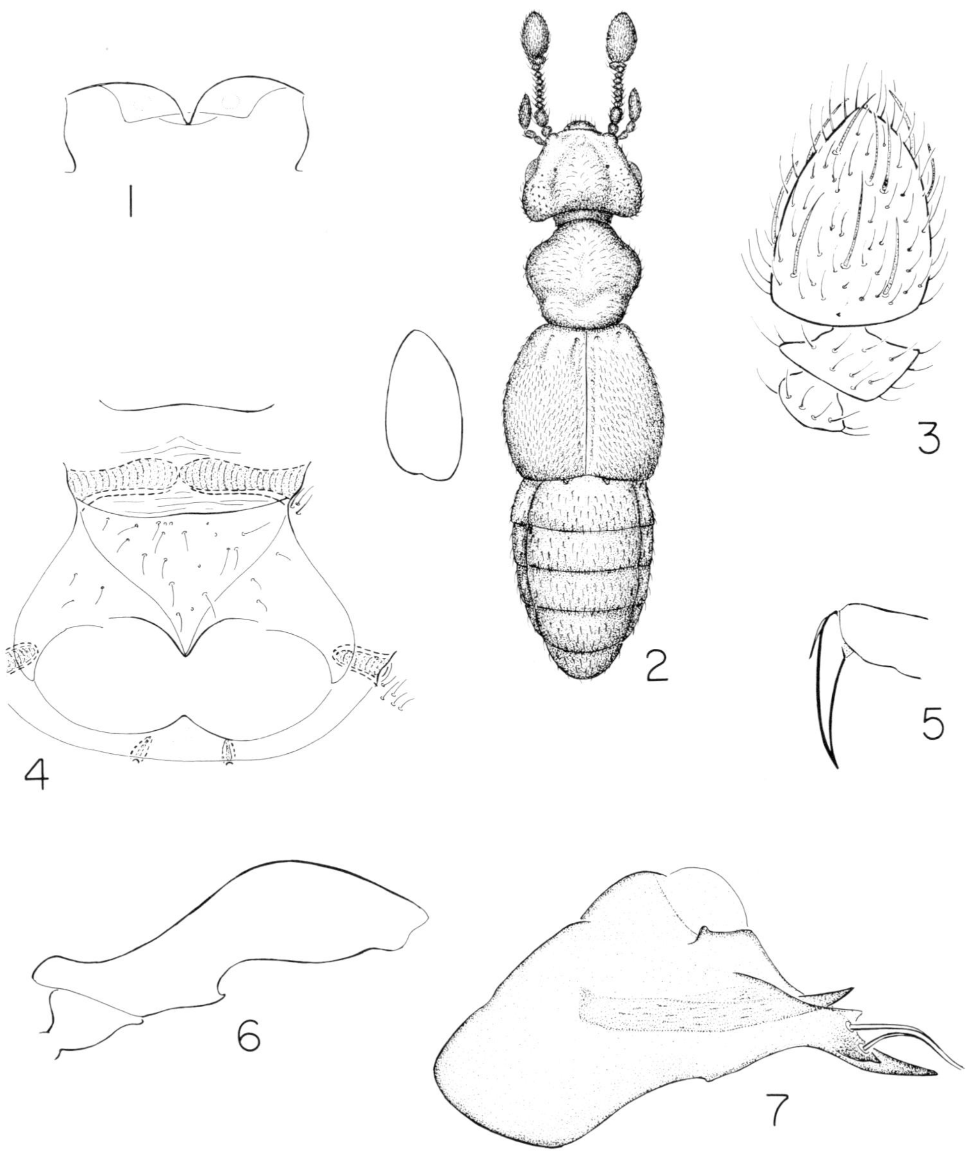

PLATE 53. *Allobrox dampfi* Fletcher. Group C. Type ♂.

Fig. 1. Prosternum.
Fig. 2. Mesosternal area.
Fig. 3. Dorsal aspect.
Fig. 4. Sensory setae of profemur.
Fig. 5. Tergite I.

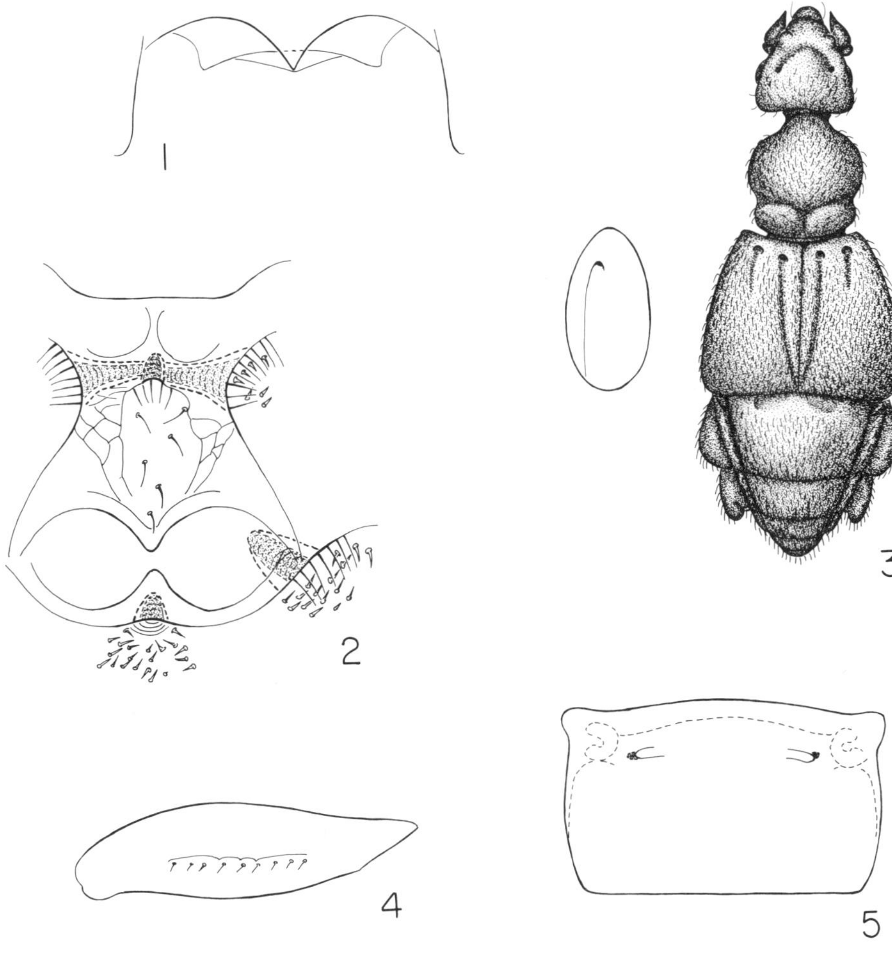

PLATE 54. *Allobrox mexicana* (Park). Group C. Type ♂.

Fig. 1. Prosternal area.
Fig. 2. Lateral margin sternite III.
Fig. 3. Lateral margin sternite IV.
Fig. 4. Antennal segments IX to XI.
Fig. 5. Tergite I.
Fig. 6. Mesosternal area.
Fig. 7. Genitalia, lateral aspect.

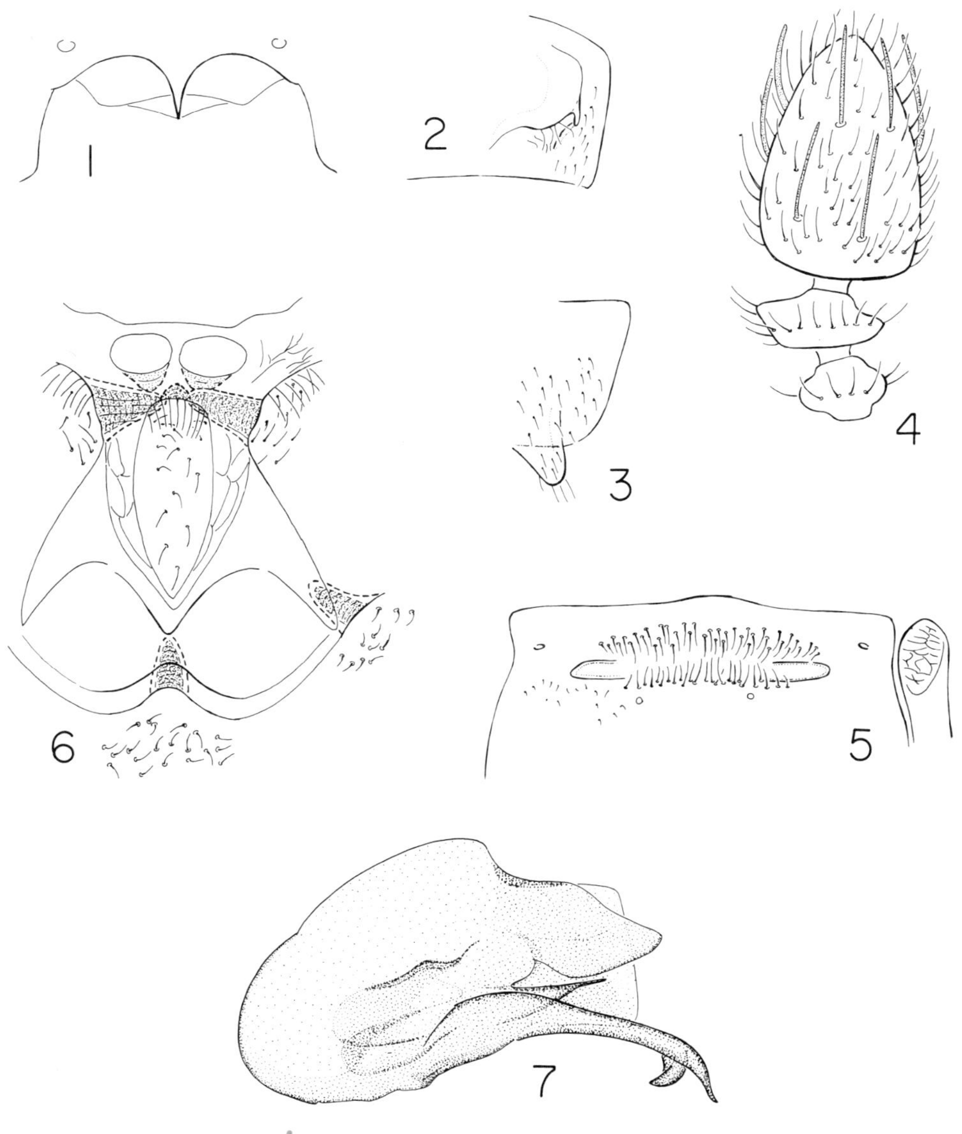

PLATE 55. *Actiastes foveicollis* (LeConte). Group C. Homotype ♂, ♀. Figs. 1-6, 8 ♂; fig. 7 ♀.

Fig. 1. Mesosternal area.
Fig. 2. Dorsal aspect.
Fig. 3. Antennal segments VII to XI.
Fig. 4. Setate modification of sternite III.
Fig. 5. Mesotarsal claws.
Fig. 6. Tergite I.
Fig. 7. Abdominal segments VIII and IX.
Fig. 8. Genitalia, lateral aspect.

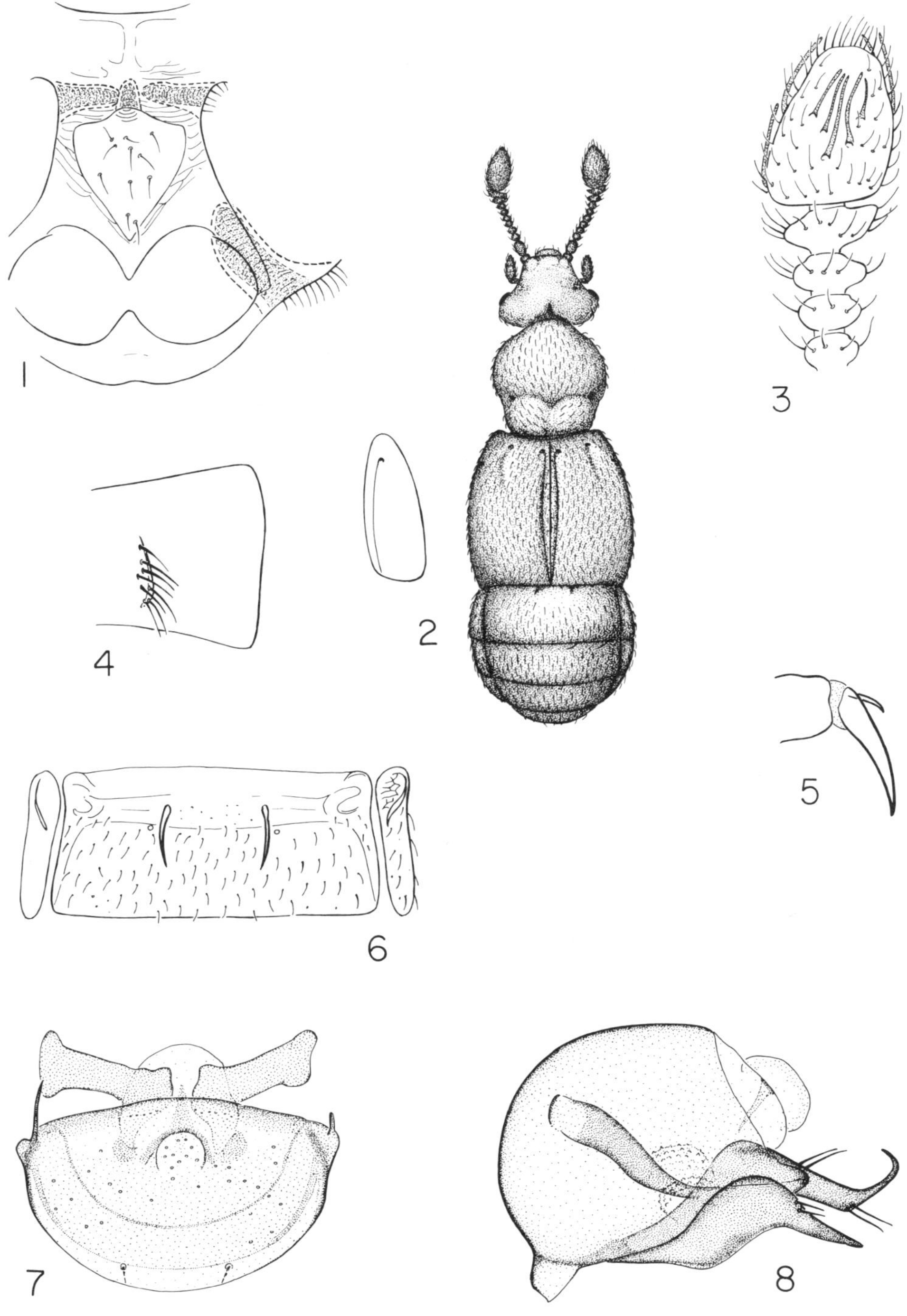

PLATE 56. *Simplona arizonica* Casey. Group C. Type ♀. Fig. 2. Homotype ♂, ♀. Figs. 1, 3, 4, 6, 8, 9 ♂; figs. 5, 7 ♀.

Fig. 1. Prosternum.
Fig. 2. Dorsal aspect.
Fig. 3. Antennal segments VII to XI.
Fig. 4. Mesosternal area.
Fig. 5. Mesotarsal claws.
Fig. 6. Sternite II, lateral margin.
Fig. 7. Abdominal segments VIII and IX.
Fig. 8. Penial plate.
Fig. 9. Genitalia, dorsal aspect.

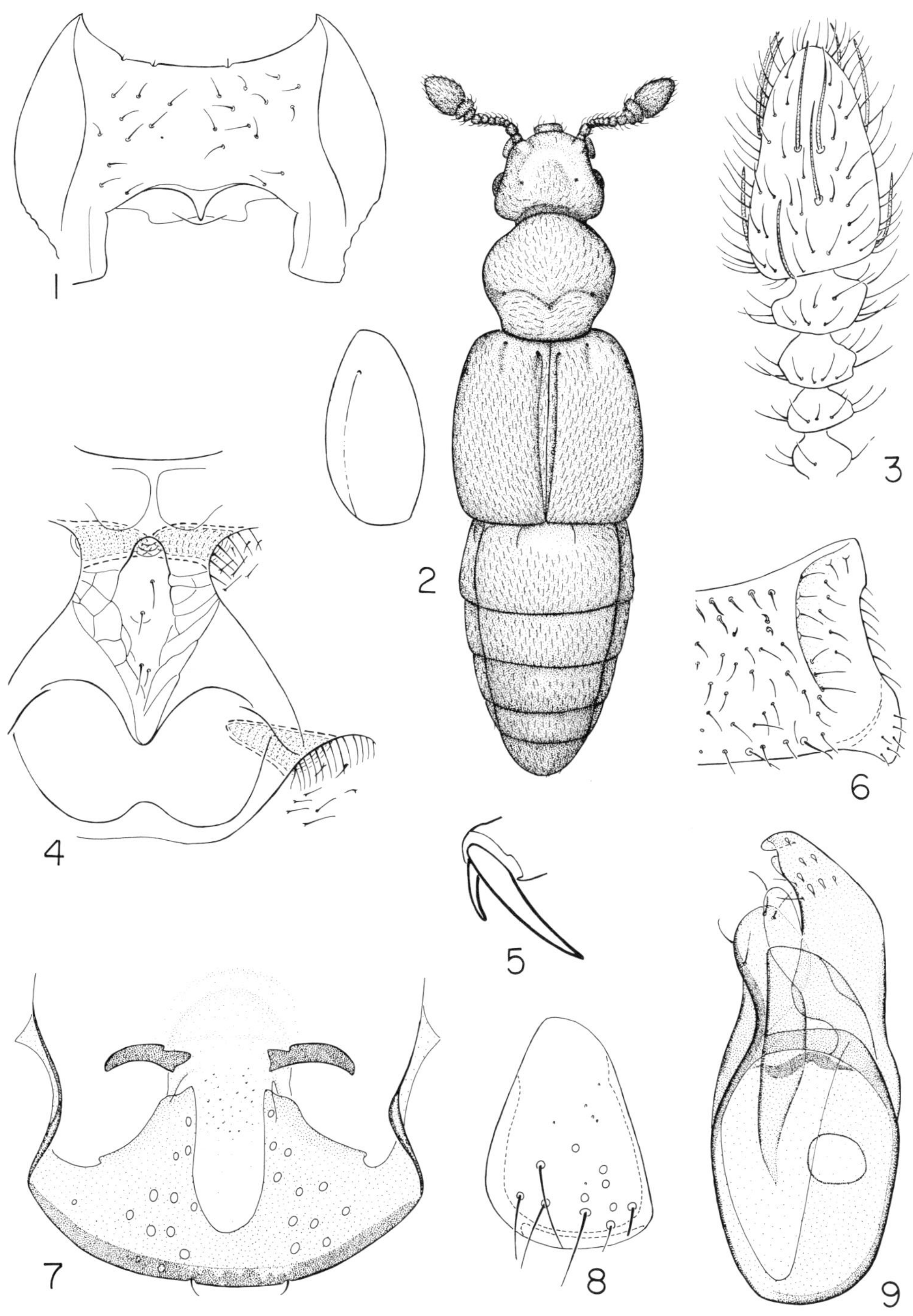

PLATE 57. *Dalmosella tenuis* Casey. Group C. Type ♀.

Fig. 1. Dorsal aspect of head.
Fig. 2. Antennal segment XI.
Fig. 3. Dorsal aspect.
Fig. 4. Prosternum.
Fig. 5. Metatarsal claws.
Fig. 6. Tergite I.
Fig. 7. Mesosternal area.
Fig. 8. Abdominal segments VIII and IX.

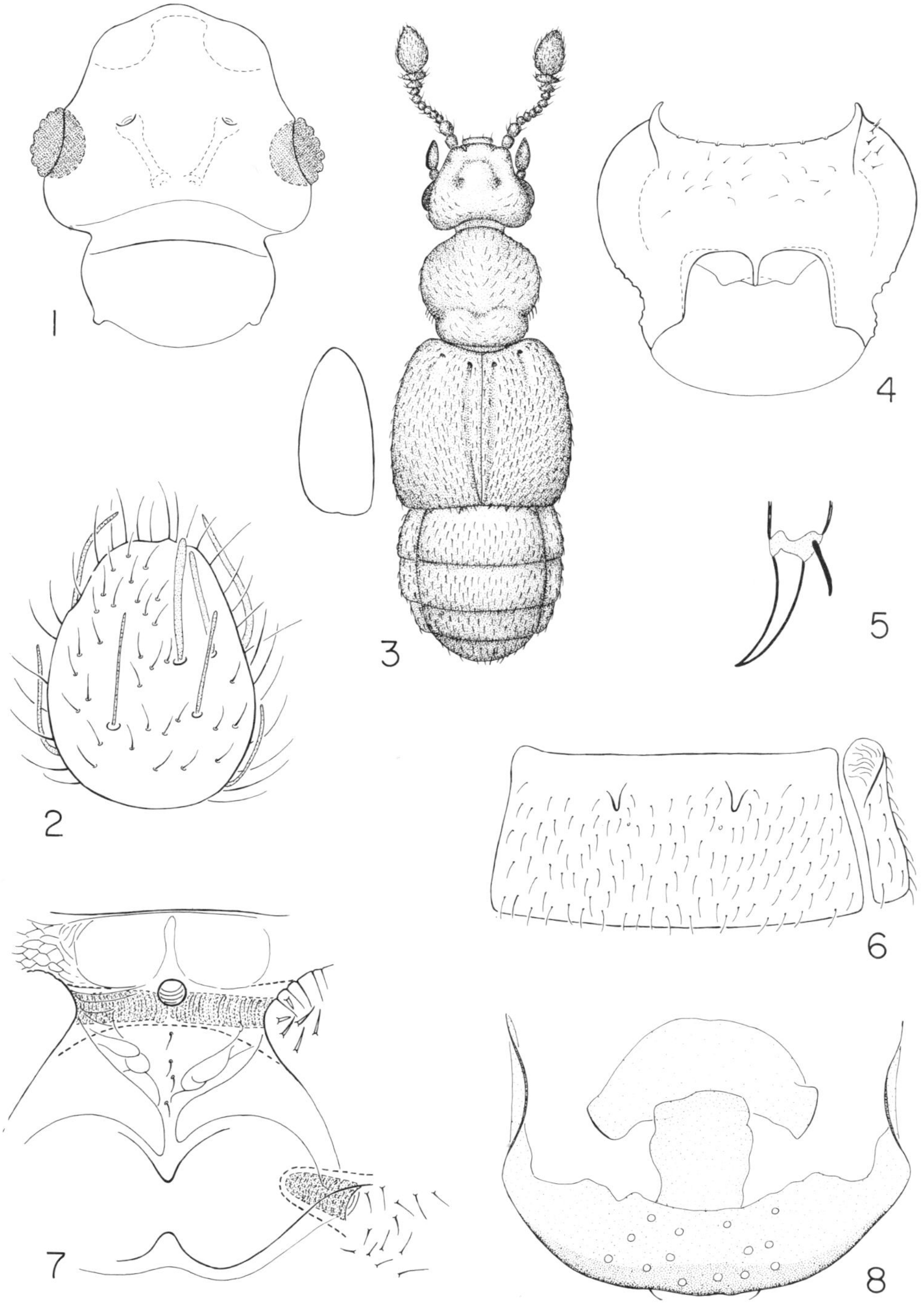

PLATE 58. *Biblomelba profunda* Park. Group C. Type ♀.

Fig. 1. Prosternum.
Fig. 2. Dorsal aspect.
Fig. 3. Antennal segments IX to XI.
Fig. 4. Mesosternal area.
Fig. 5. Metacoxae.
Fig. 6. Profemur.

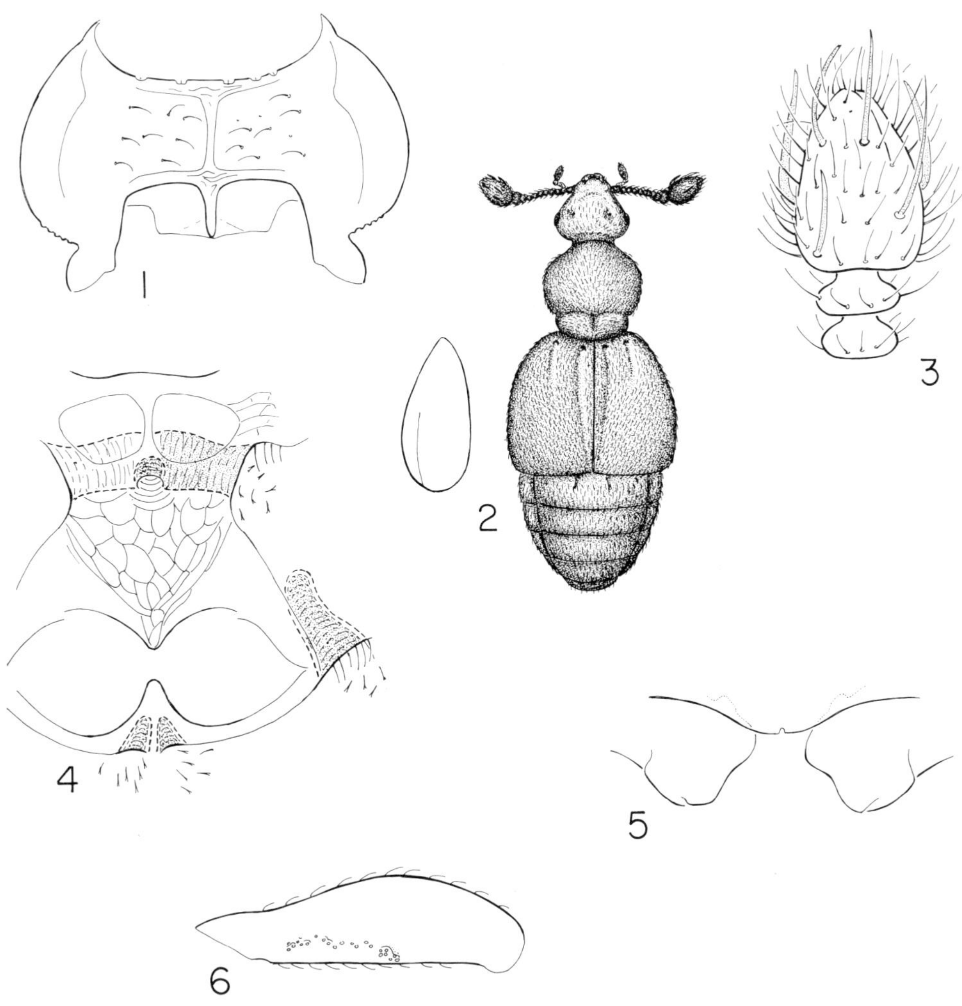

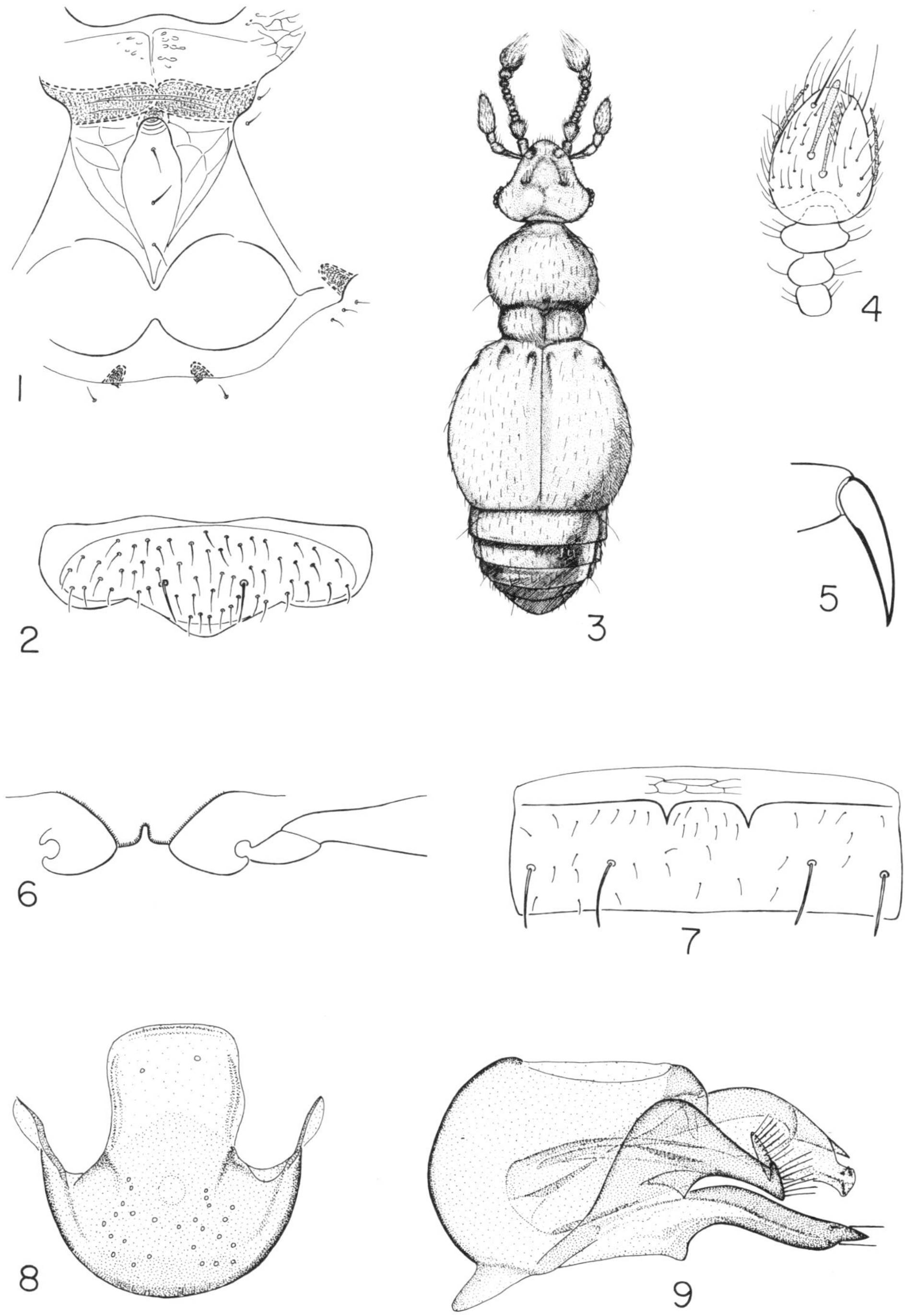

PLATE 60. *Trimioarcus incisurus* Park. Group C. Type ♂.

Fig. 1. Prosternal area.
Fig. 2. Dorsal aspect.
Fig. 3. Antennal segments IX to XI.
Fig. 4. Mesosternal claws.
Fig. 5. Profemur.
Fig. 6. Metatarsal claw.
Fig. 7. Genitalia, dorsal aspect.

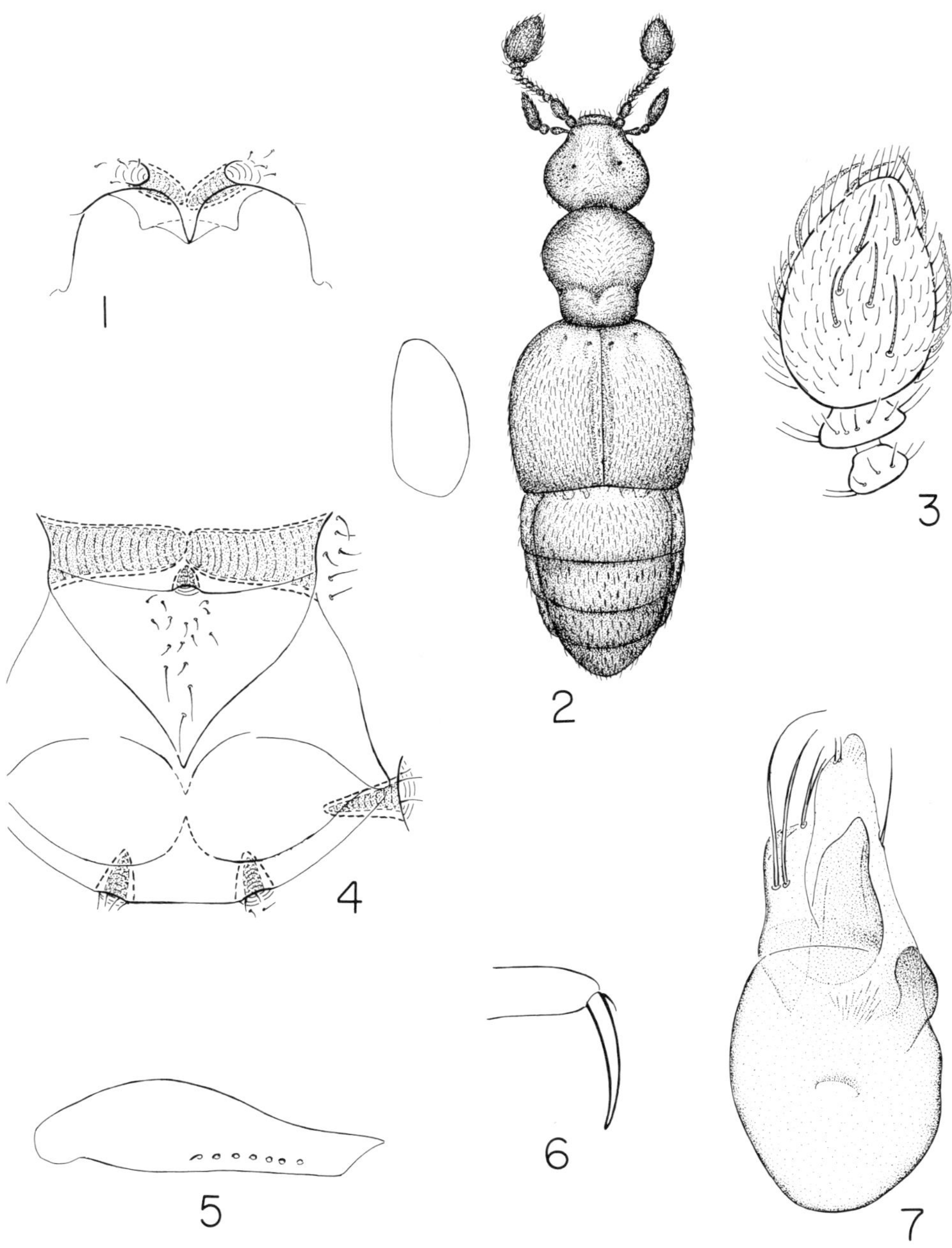

PLATE 61. *Dalmosanus inoculus* Park. Group C. Type ♀. Fig. 1. *Dalmosanus acuta* (Park). Type ♂. Figs. 2-6.

Fig. 1. Dorsal aspect.
Fig. 2. Antennal segments IX to XI.
Fig. 3. Dorsal aspect.
Fig. 4. Metatarsal claw.
Fig. 5. Mesosternal area.
Fig. 6. Male genitalia, dorsal aspect.

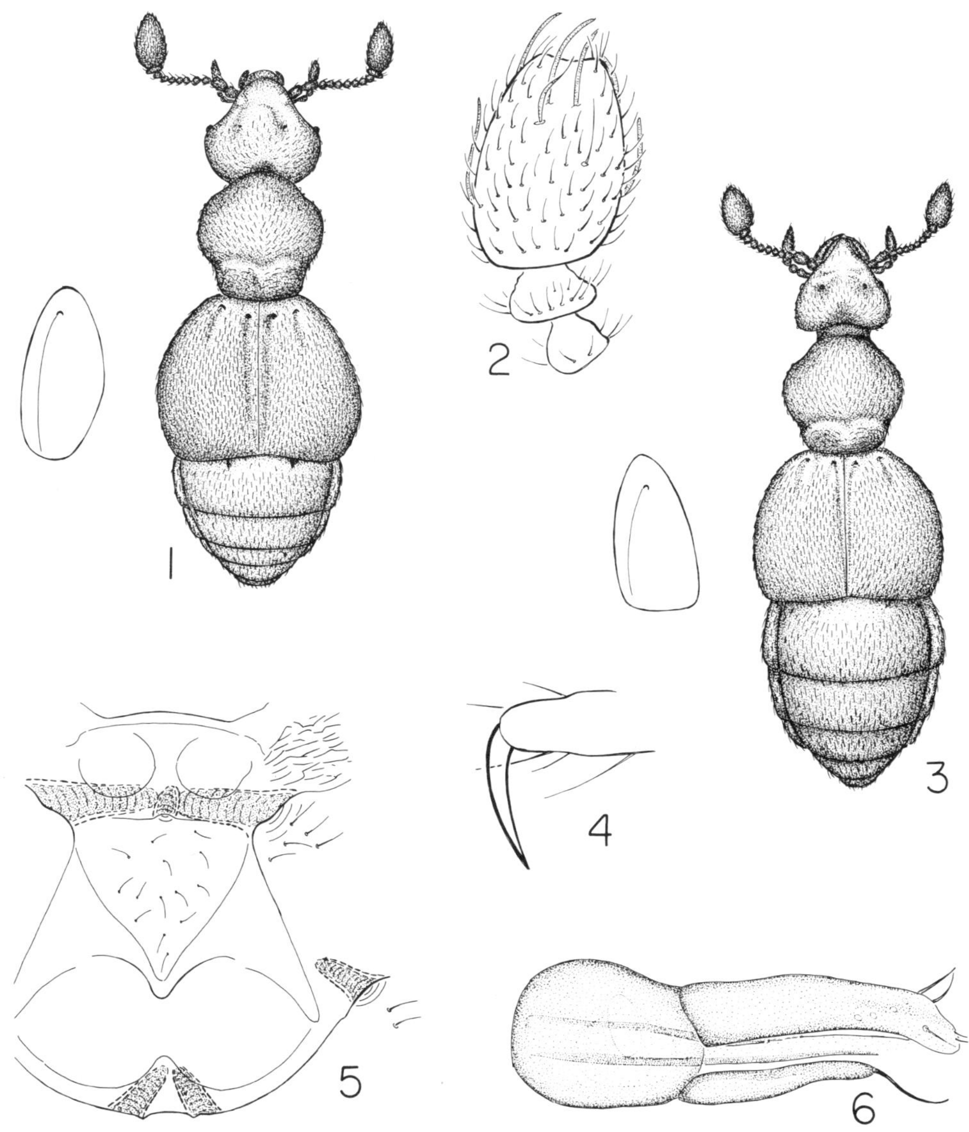

PLATE 62. *Minibi zonula* (Park). Group C. Paratype ♂.

    Fig. 1. Procoxal area.
    Fig. 2. Dorsal aspect.
    Fig. 3. Antennal segments IX to XI.
    Fig. 4. Mesotrochanter.
    Fig. 5. Mesosternal area.
    Fig. 6. Genitalia, lateral aspect.

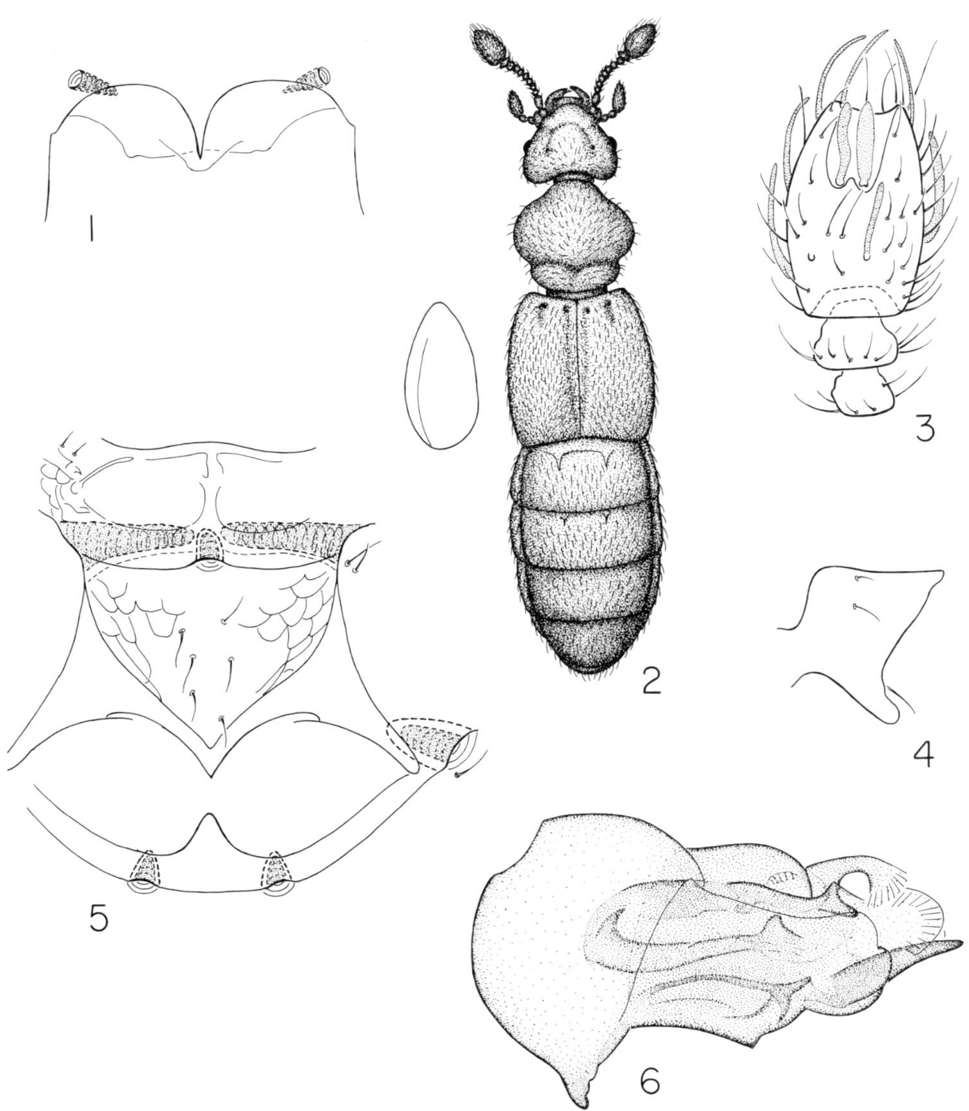

PLATE 63. *Allomelba antennata* Park. Group C. Type ♂.

Fig. 1. Prosternum.
Fig. 2. Metatarsal claw.
Fig. 3. Dorsal aspect.
Fig. 4. Antennal segments IX to XI.
Fig. 5. Mesosternal area.
Fig. 6. Mesotrochanter.
Fig. 7. Genitalia, lateral aspect.

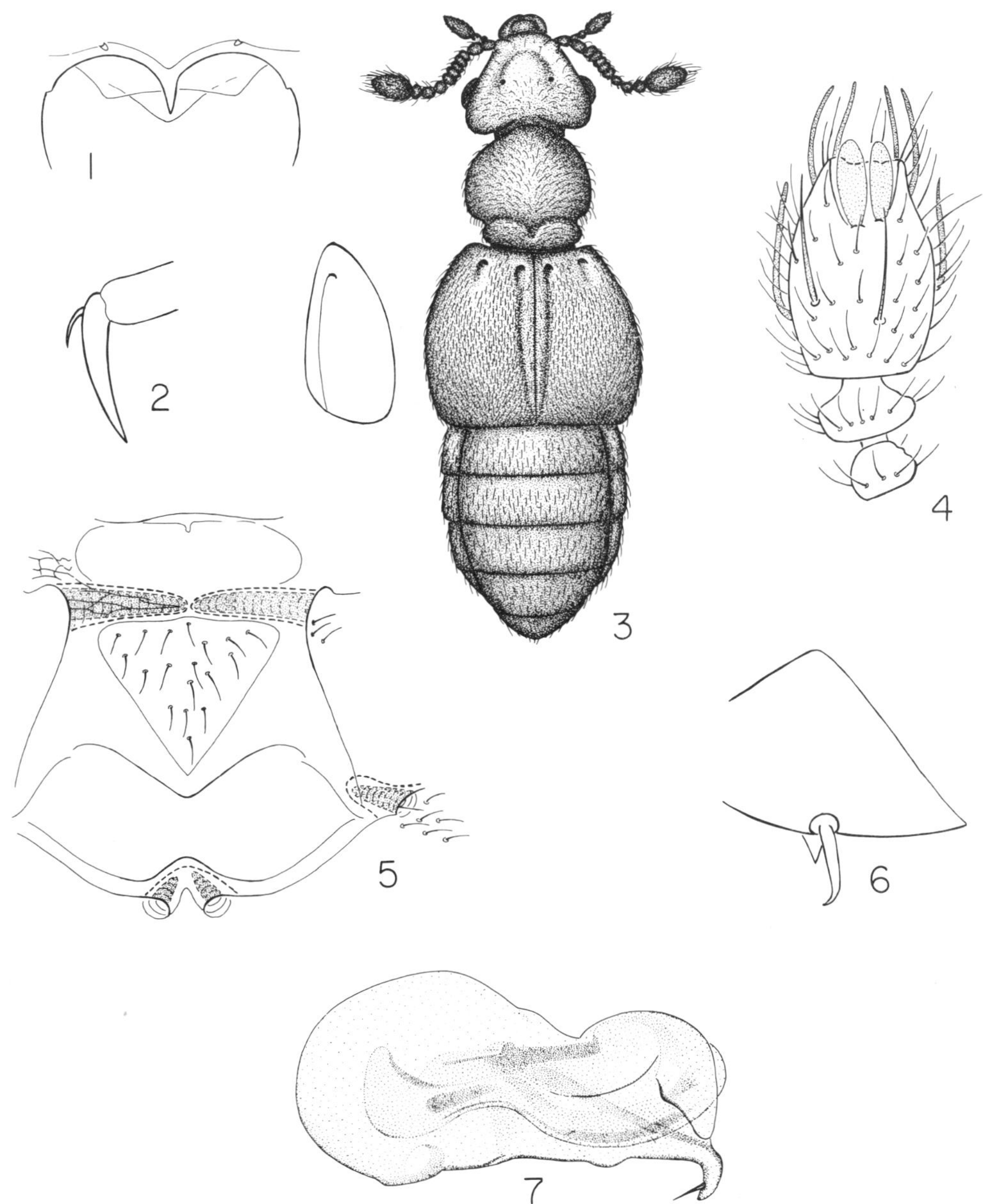

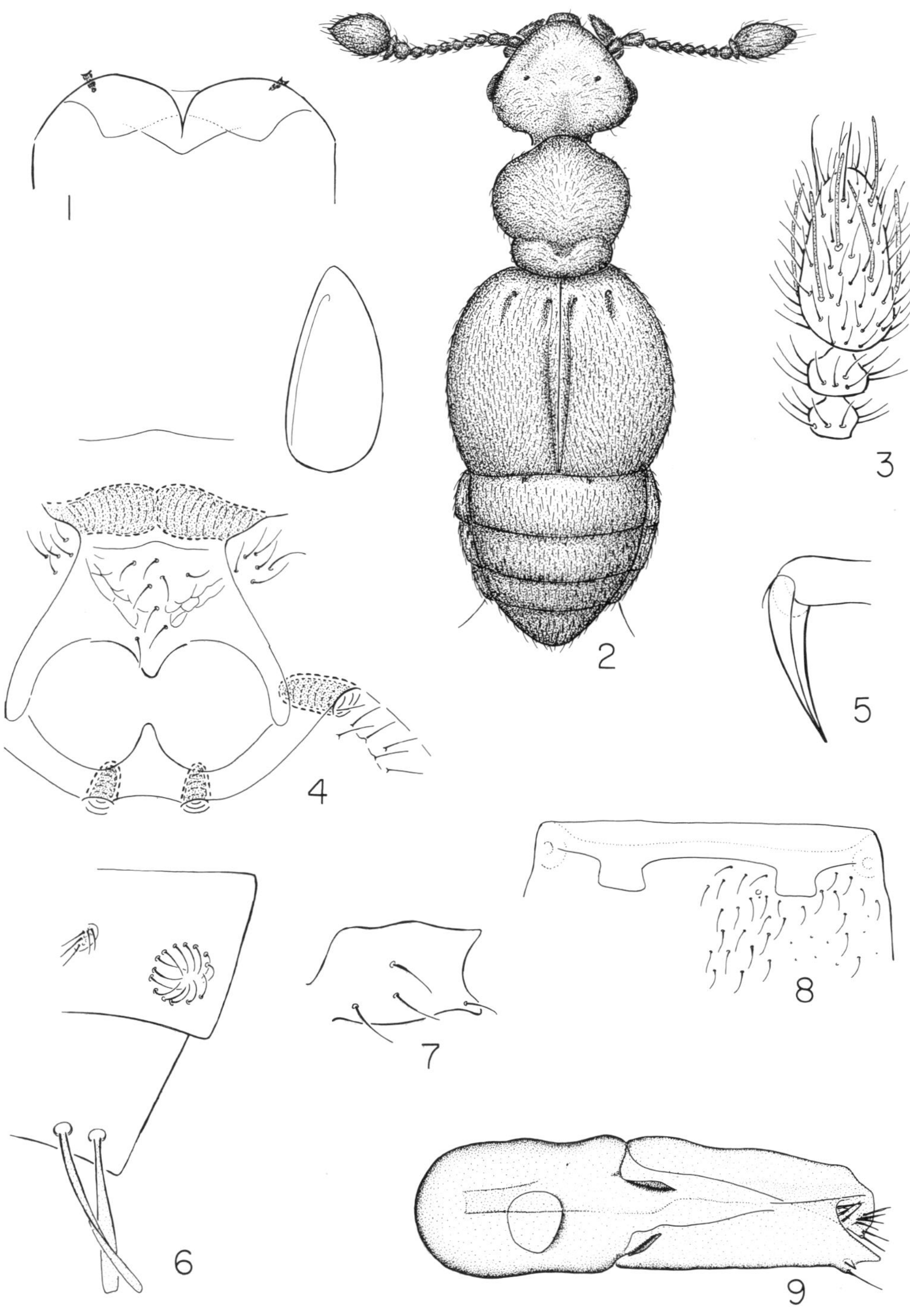

PLATE 65. *Trimiodina concolor* (Sharp). Group C. Lectotype ♂. Figs. 1, 3-9. "Type"
at MNHN. Fig. 2.

Fig. 1. Prosternum.
Fig. 2. Dorsal aspect.
Fig. 3. Antennal segments IX to XI.
Fig. 4. Mesosternal area.
Fig. 5. Metatrochanter.
Fig. 6. Mesotarsal claws.
Fig. 7. Lateral margins sternites III and IV.
Fig. 8. Sternite VI.
Fig. 9. Genitalia, dorsal aspect (apex of median lobe may be incomplete).

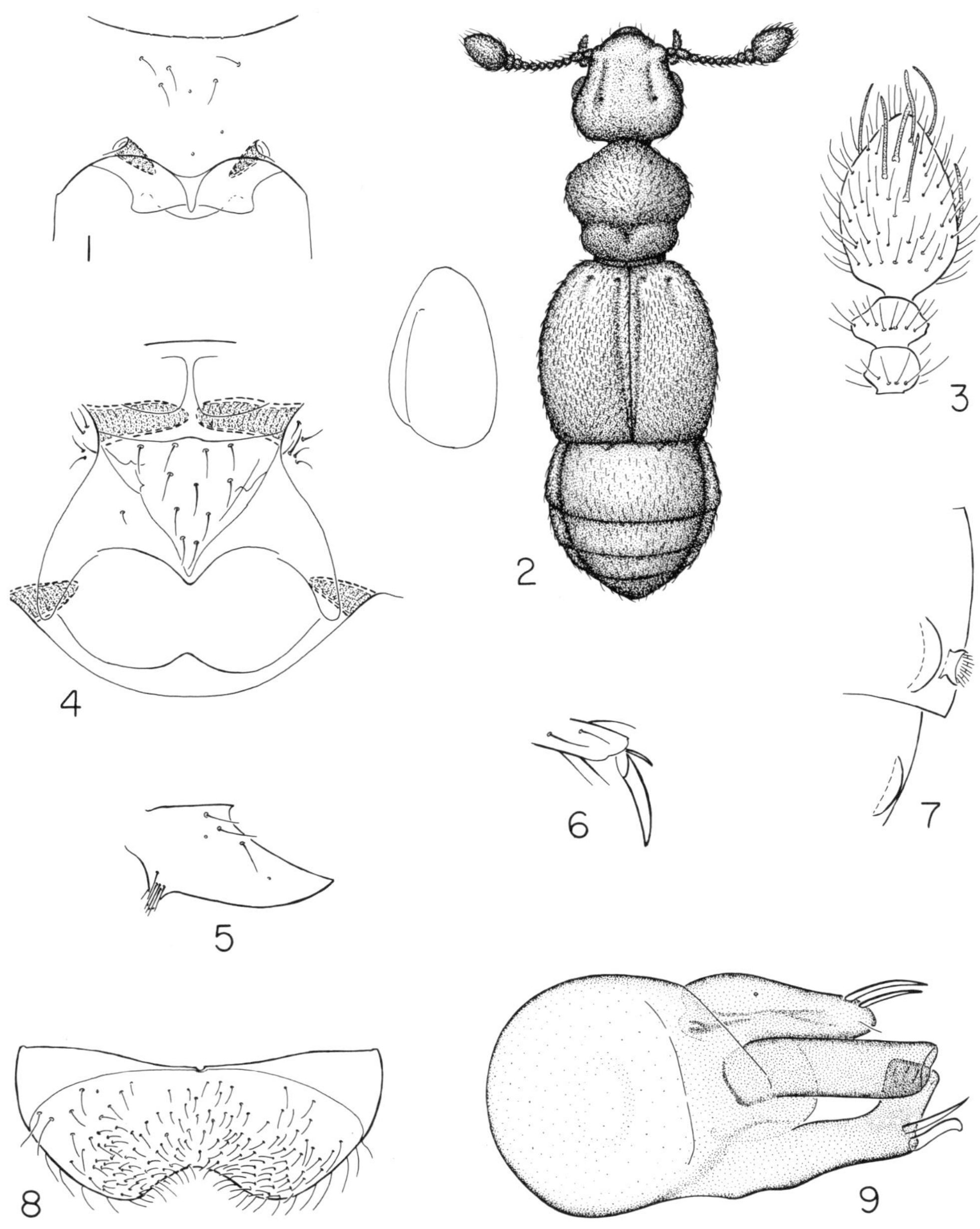

PLATE 66. *Quadrelba parmata* (Reitter). Group C. ♂.

Fig. 1. Prosternum.
Fig. 2. Dorsal aspect.
Fig. 3. Antennal segments IX to XI.
Fig. 4. Mesosternal area.
Fig. 5. Mesotrochanter.
Fig. 6. Mesotarsal claws.
Fig. 7. Genitalia, dorsal aspect.

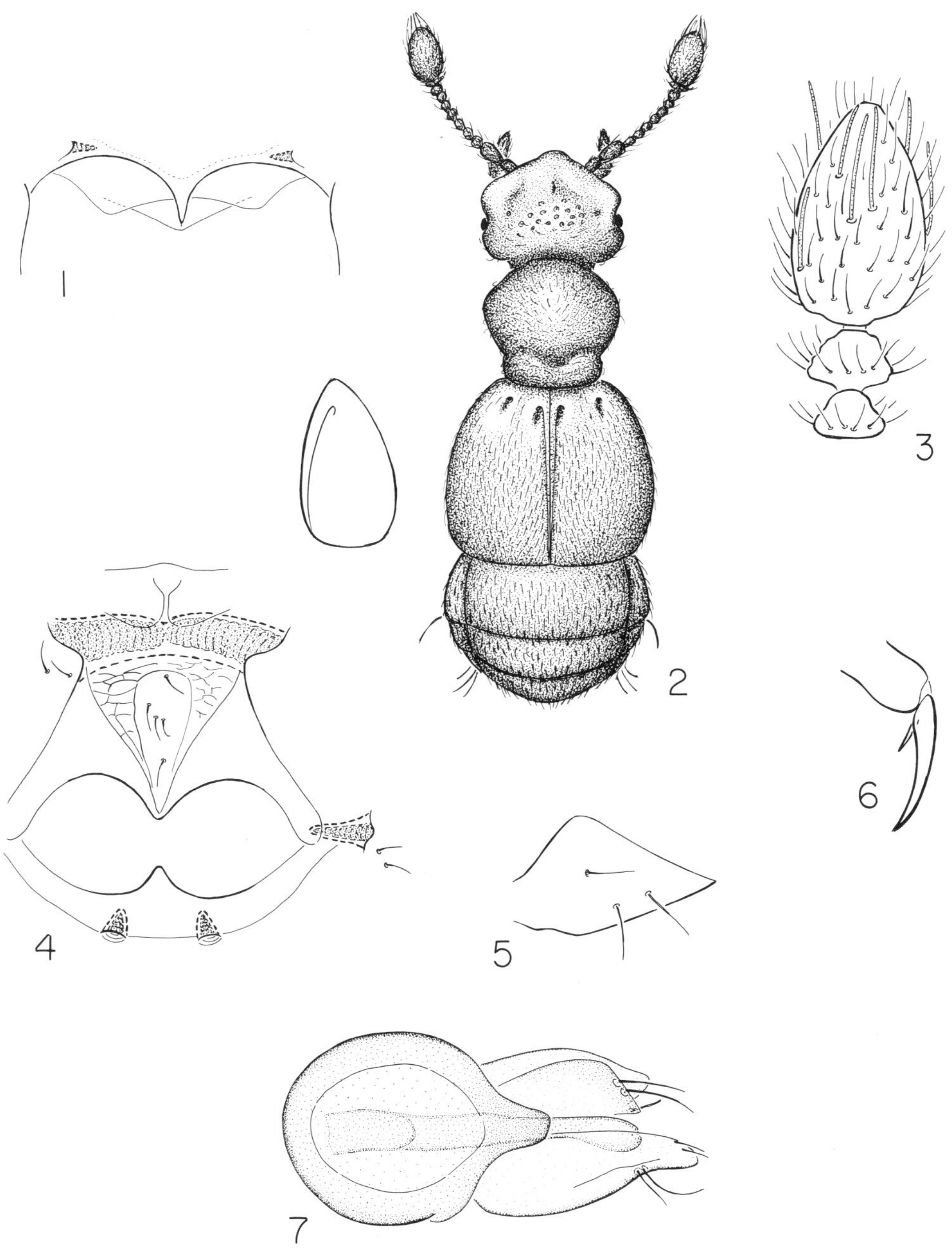

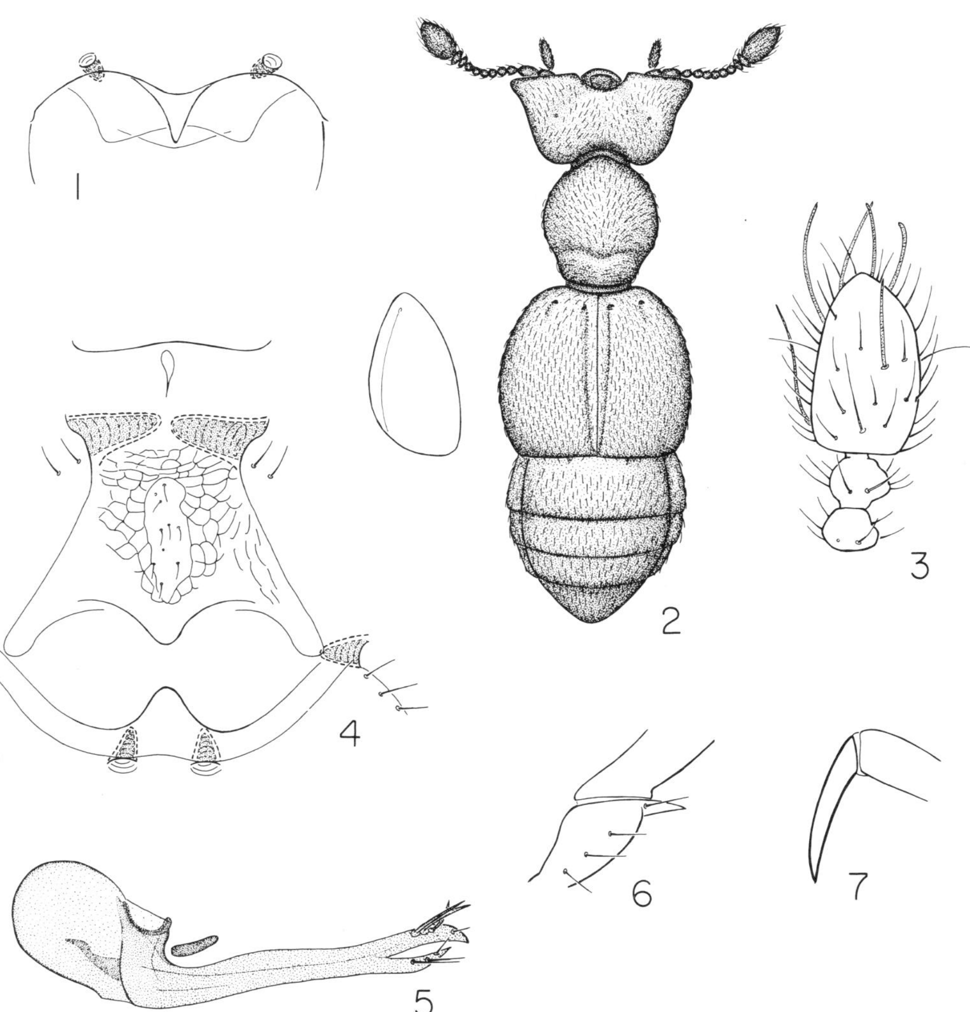

PLATE 68. *Trimiovillus bahamicus* Park. Group C. Type ♀.

Fig. 1. Prosternum.
Fig. 2. Antennal segments IX to XI.
Fig. 3. Dorsal aspect.
Fig. 4. Metatarsal claws.
Fig. 5. Mesosternal area.
Fig. 6. Profemur.
Fig. 7. Abdominal segments VIII and IX.

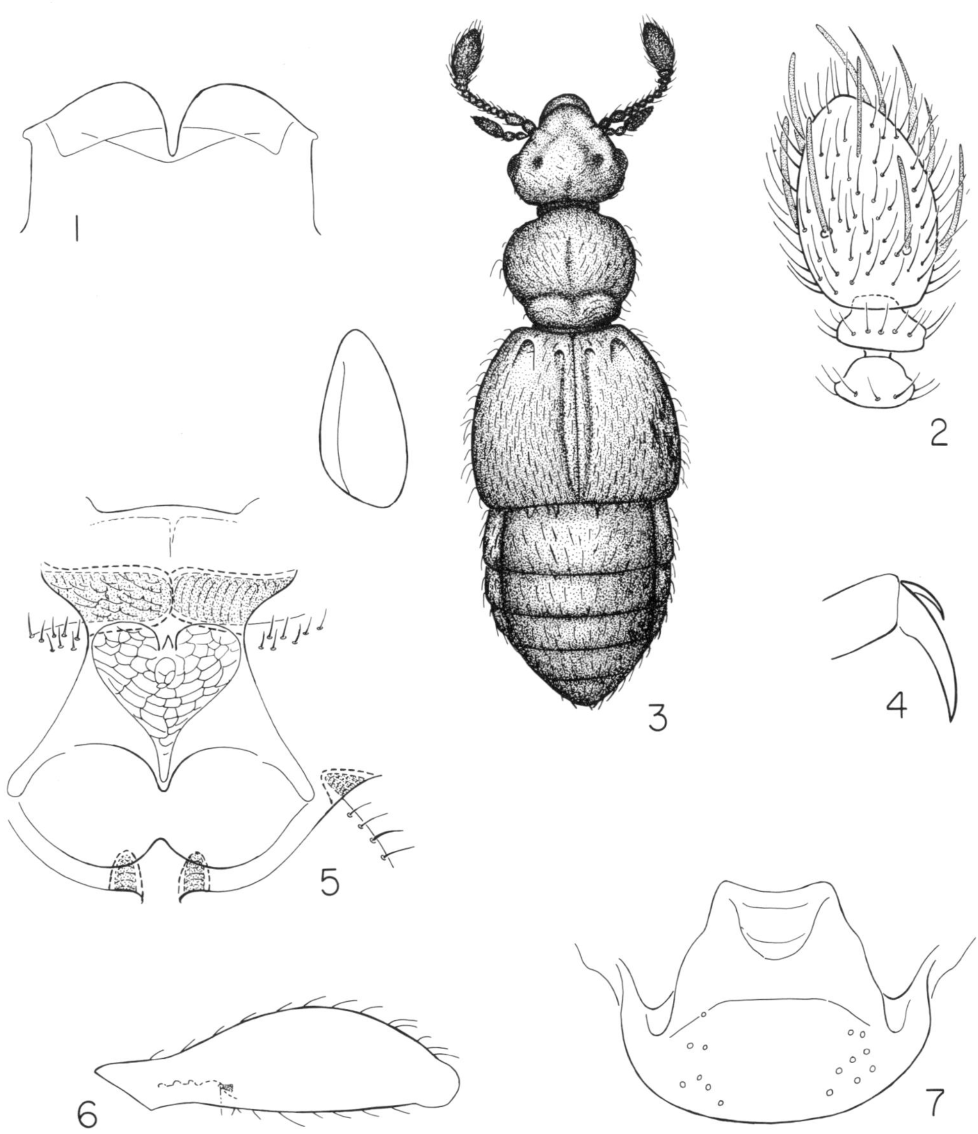

PLATE 69. *Trimiomelba convexula* (LeConte). Group C. Type ♀.

Fig. 1. Prosternal area.
Fig. 2. Dorsal aspect.
Fig. 3. Antennal segments IX to XI.
Fig. 4. Mesosternal area.
Fig. 5. Metatarsal claws.
Fig. 6. Profemur.
Fig. 7. Tergite I.

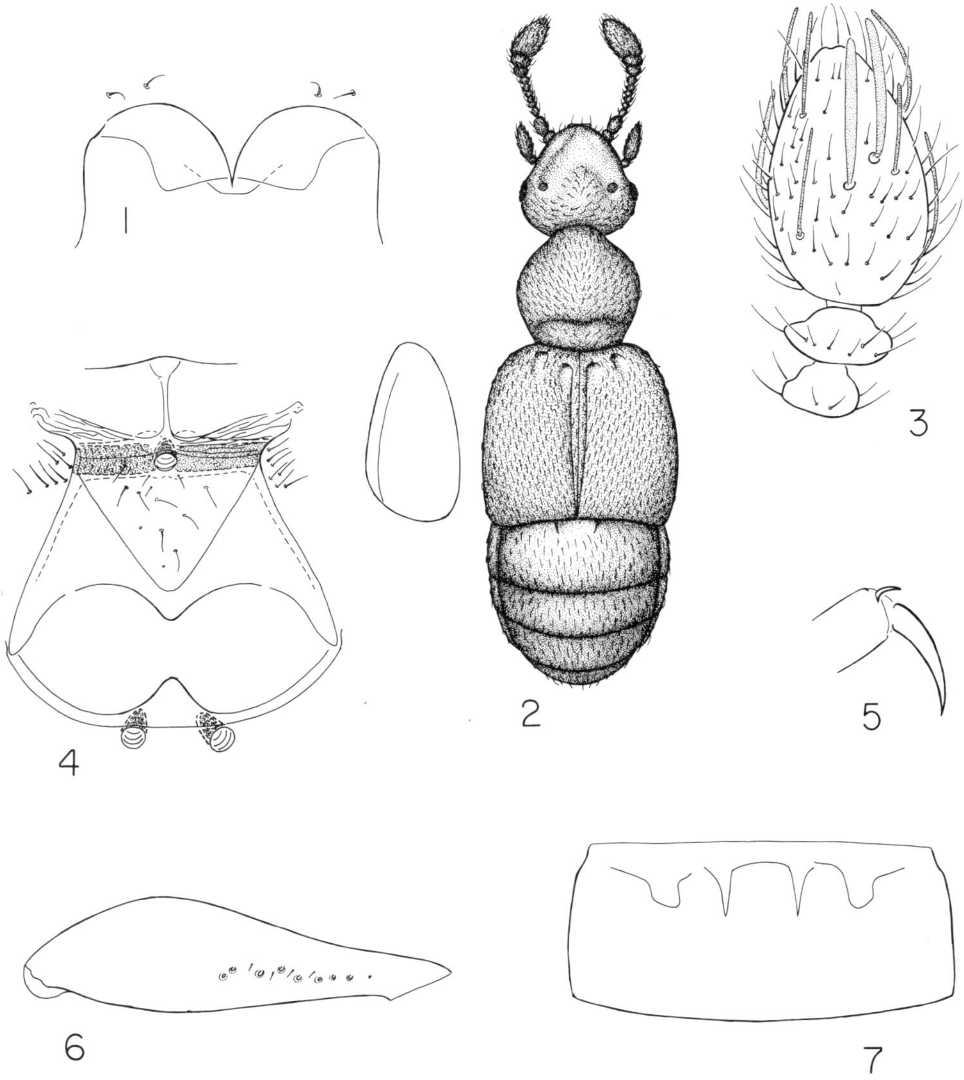
1
2
3
4
5
6
7

PLATE 70. *Allotrimium michoacanensis* Park. Group C. Type ♂.

Fig. 1. Anterior margin procoxal cavities.
Fig. 2. Dorsal aspect.
Fig. 3. Antennal segments IX to XI.
Fig. 4. Mesosternal area.
Fig. 5. Mesotrochanter, femur and tibia, ventral.
Fig. 6. Metatarsal claws.
Fig. 7. Genitalia, dorsal aspect.

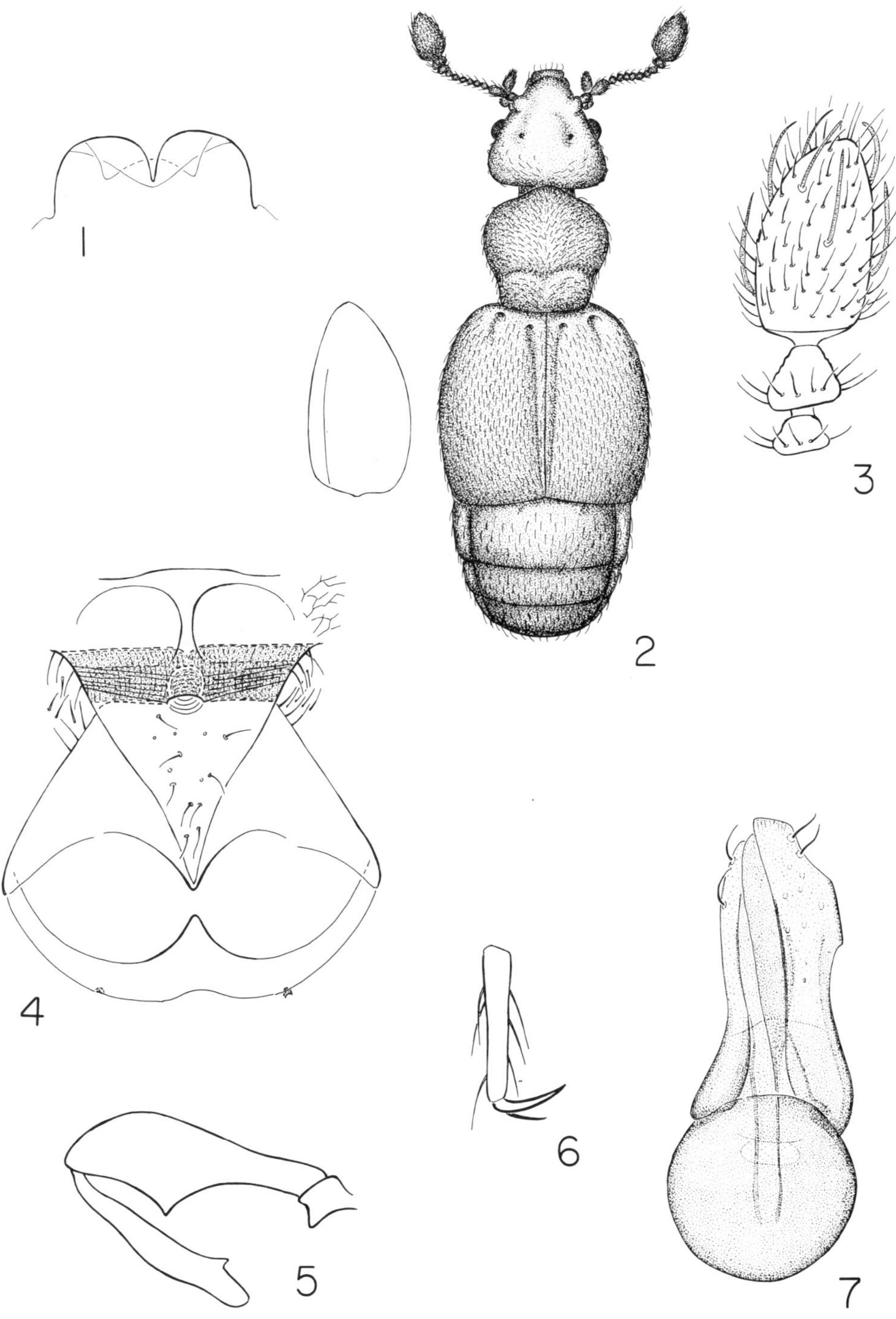

PLATE 71. *Melba thoracicum* (Brendel). Group C. Lectotype ♀.

Fig. 1. Prosternal area.
Fig. 2. Metatarsal claws.
Fig. 3. Dorsal aspect.
Fig. 4. Antennal segments IX to XI.
Fig. 5. Mesosternal area.
Fig. 6. Profemur.

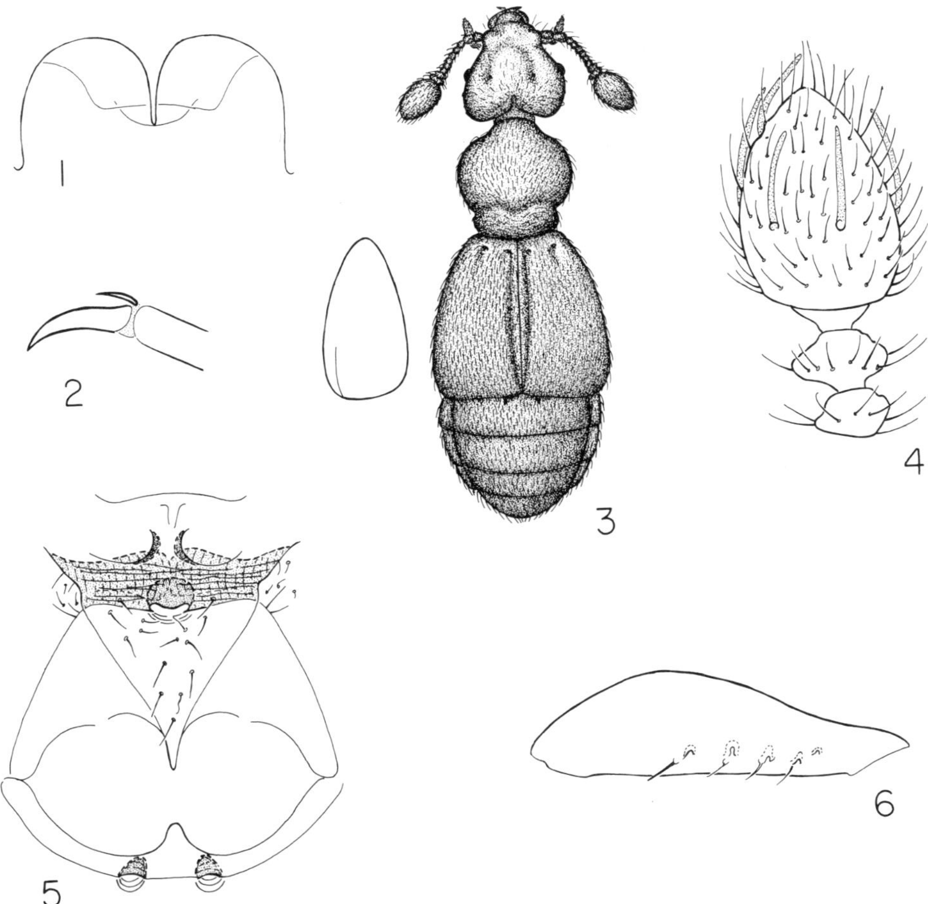

PLATE 72. *Zonaira trilinea* Grigarick and Schuster. Group C. Type ♂, paratype ♀.
Figs. 1, 3-8, 10, 11 ♂; figs. 2, 9 ♀.

Fig. 1. Prosternal area.
Fig. 2. Dorsal aspect.
Fig. 3. Antennal segments IX to XI.
Fig. 4. Tarsal claws.
Fig. 5. Mesosternal area.
Fig. 6. Modification on lateral margin of sternite III.
Fig. 7. Mesotibia and mesotrochanter.
Fig. 8. Mesotibia.
Fig. 9. Abdominal segments VIII and IX.
Fig. 10. Metacoxa.
Fig. 11. Genitalia, dorsal aspect.

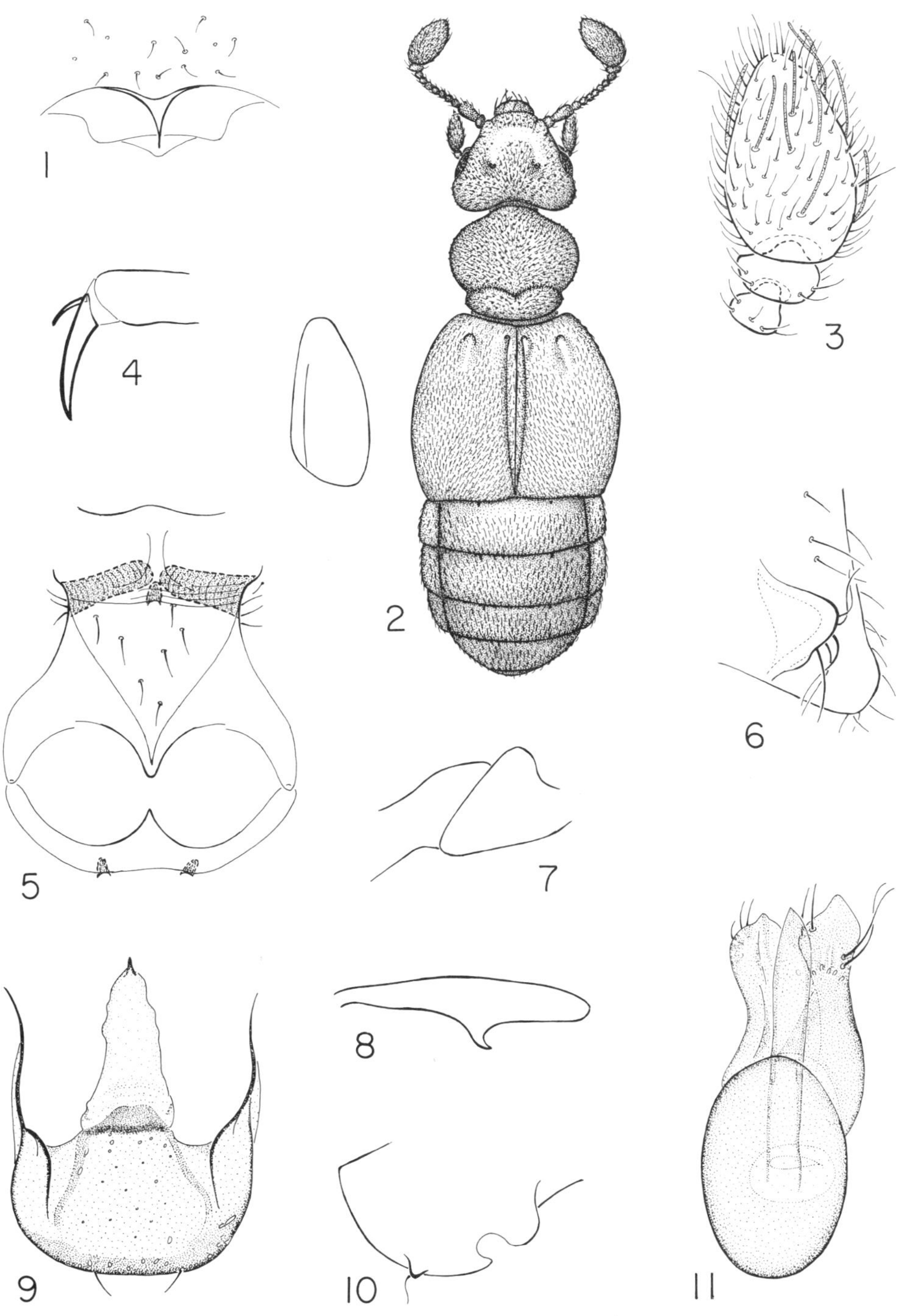

PLATE 73. *Perimelba granulosa* Park. Group C. Type ♀.

Fig. 1. Prosternum.
Fig. 2. Dorsal aspect.
Fig. 3. Antennal segments IX to XI.
Fig. 4. Mesosternal area.
Fig. 5. Profemur.
Fig. 6. Tergites I to III.

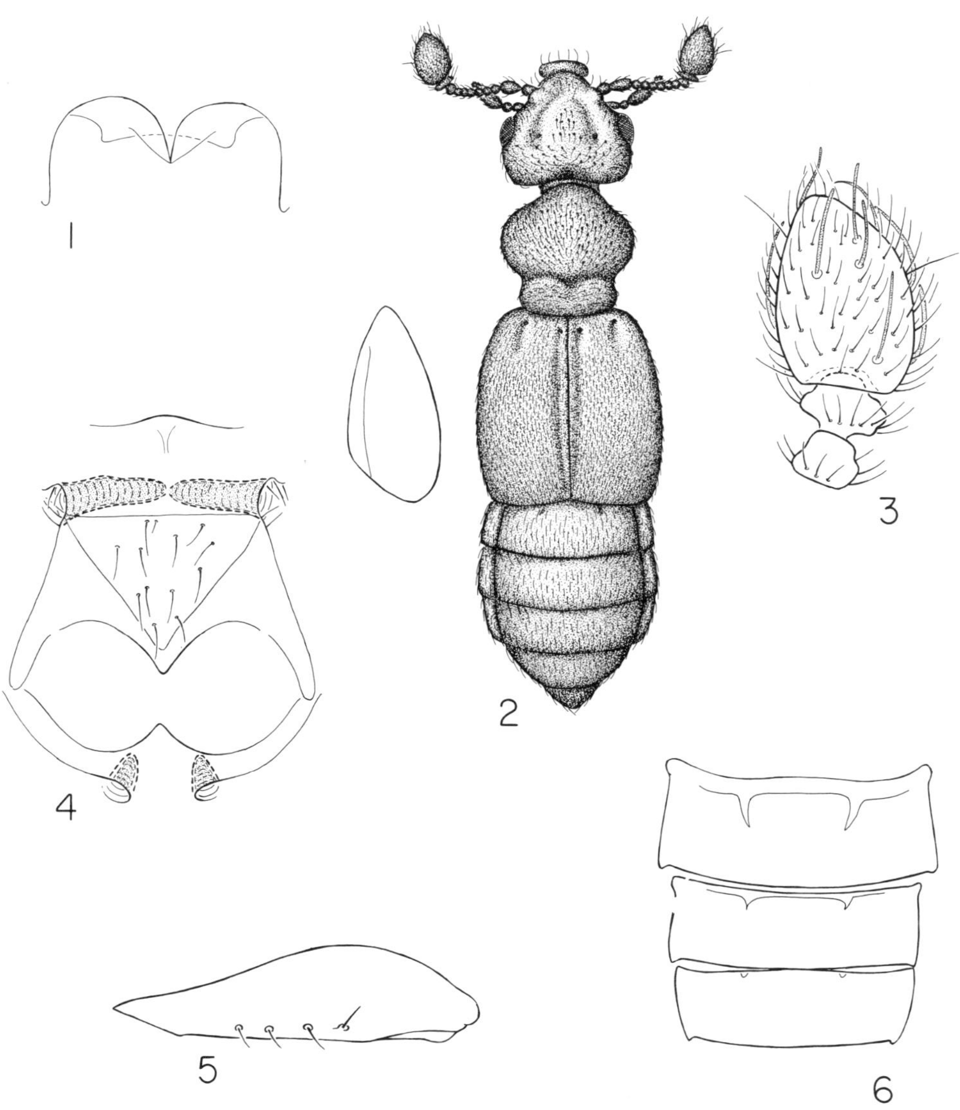
1
2
3
4
5
6

PLATE 74. *Hanfordia absoluta* Park. Group C. Homoparatopotype ♂.

Fig. 1. Prosternum.
Fig. 2. Mesotarsal claw.
Fig. 3. Dorsal aspect.
Fig. 4. Antennal segments IX to XI.
Fig. 5. Mesosternal area.
Fig. 6. Sternite VI.
Fig. 7. Mesotibia and mesofemur.
Fig. 8. Genitalia, dorsal aspect.

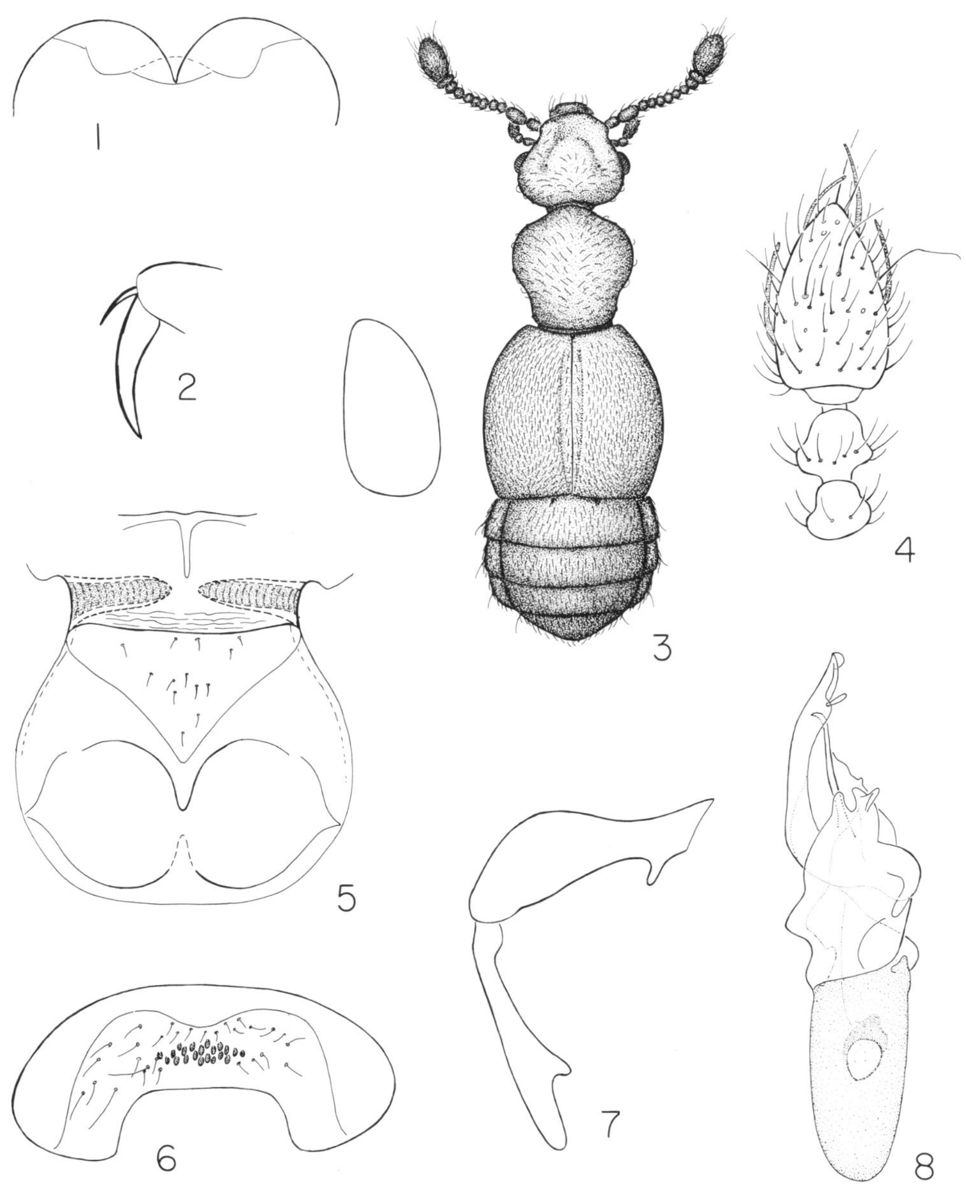

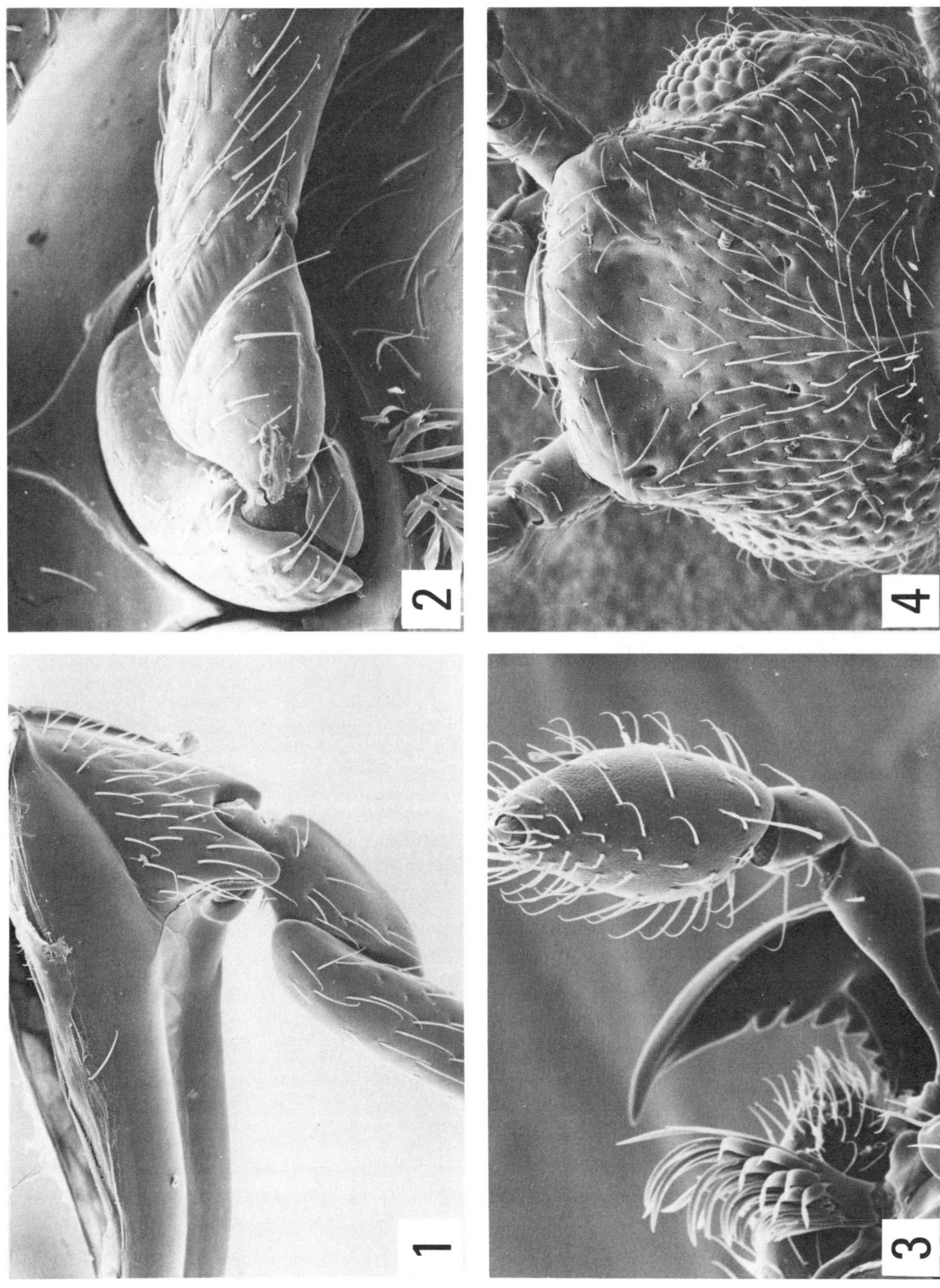

PLATE 76.

Fig. 1. Mesosternal area, external view, *Oropus castaneus*, 325X.
Fig. 2. Mesosternal area, internal view, *O. castaneus*, 367X.

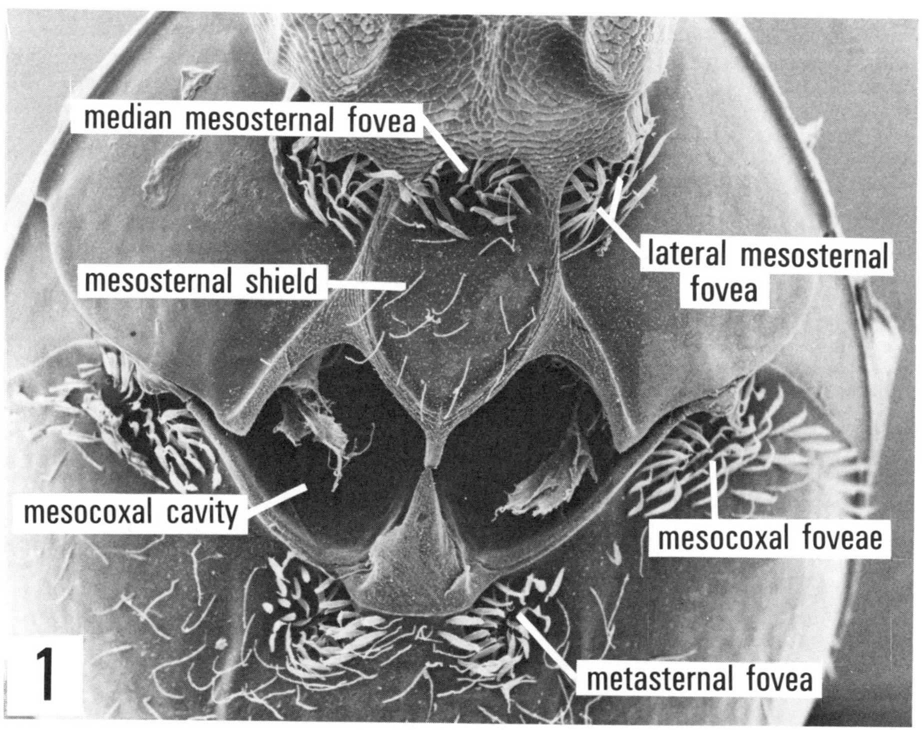

median mesosternal fovea
lateral mesosternal fovea
mesosternal shield
mesocoxal cavity
mesocoxal foveae
metasternal fovea
1

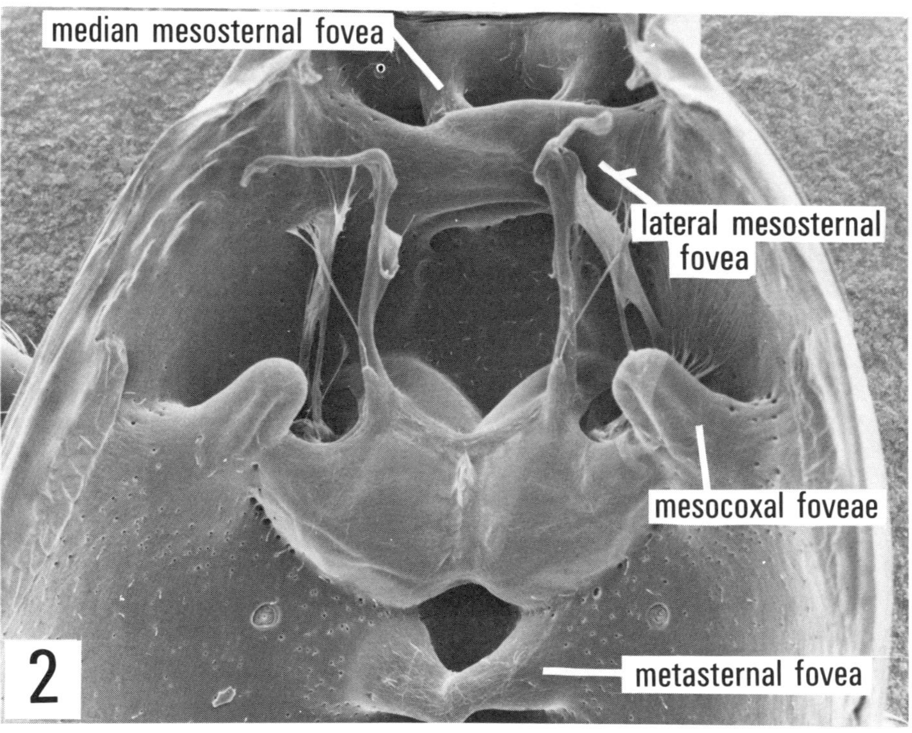

median mesosternal fovea
lateral mesosternal fovea
mesocoxal foveae
metasternal fovea
2

PLATE 77.

Fig. 1. Mesosternal area, *Rhexius sharpi*, 538X.
Fig. 2. Mesosternal area, *Allobrox* sp., 736X.
Fig. 3. Mesosternal area, *Pygmactium steevesi*, 978X.
Fig. 4. Mesosternal area, *Melba* sp., 960X.

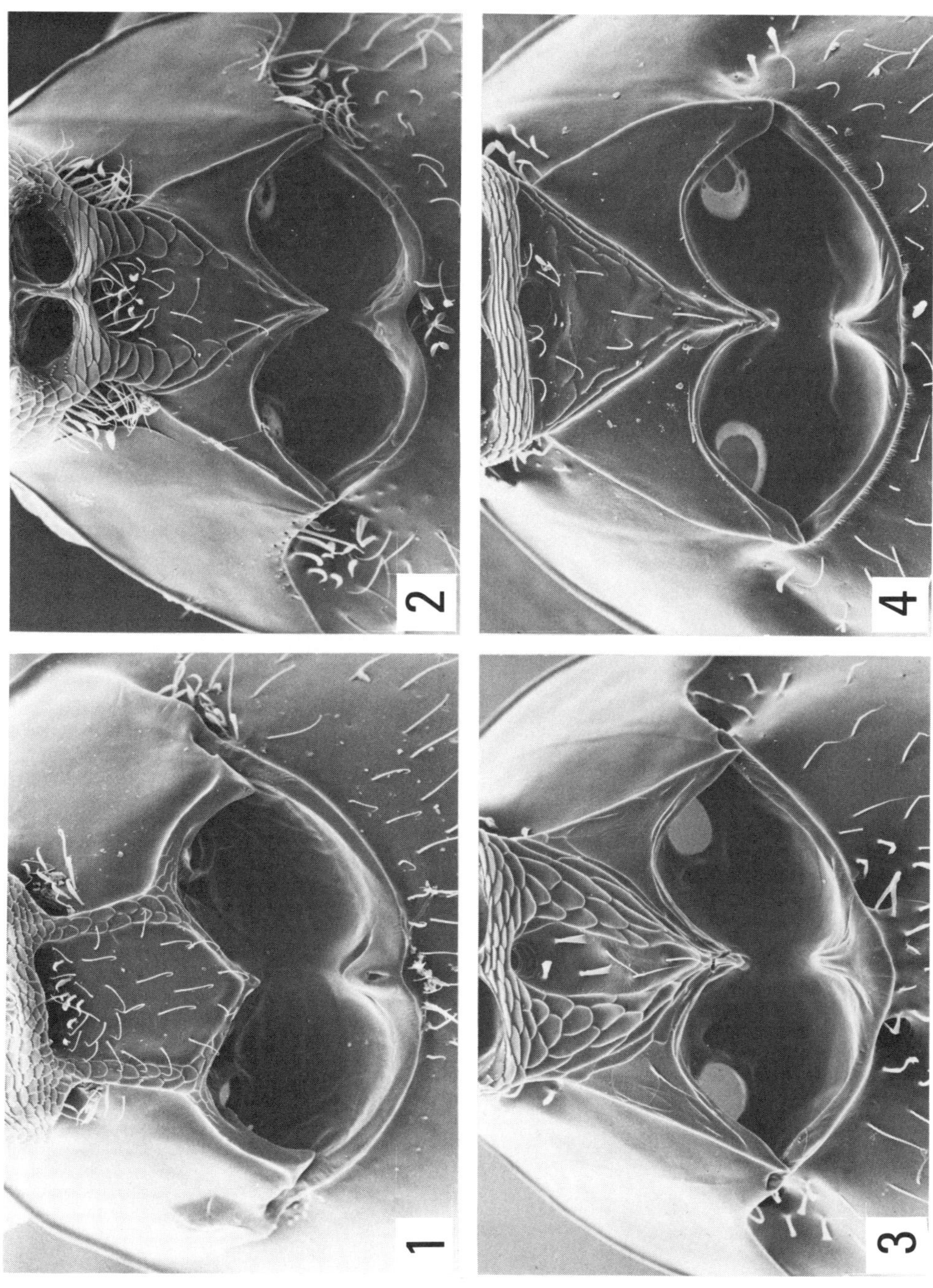

PLATE 78.

Fig. 1. Metatarsal claws, *Tetrascapha dasycerca*, 2500X.
Fig. 2. Mesotarsal claws, *Pygmactium steevesi*, 9890X.
Fig. 3. Sensory pits, profemur, *Simplona arizonica*, 5565X.
Fig. 4. Sensory pit, profemur, *Melba* sp., 12,380X.

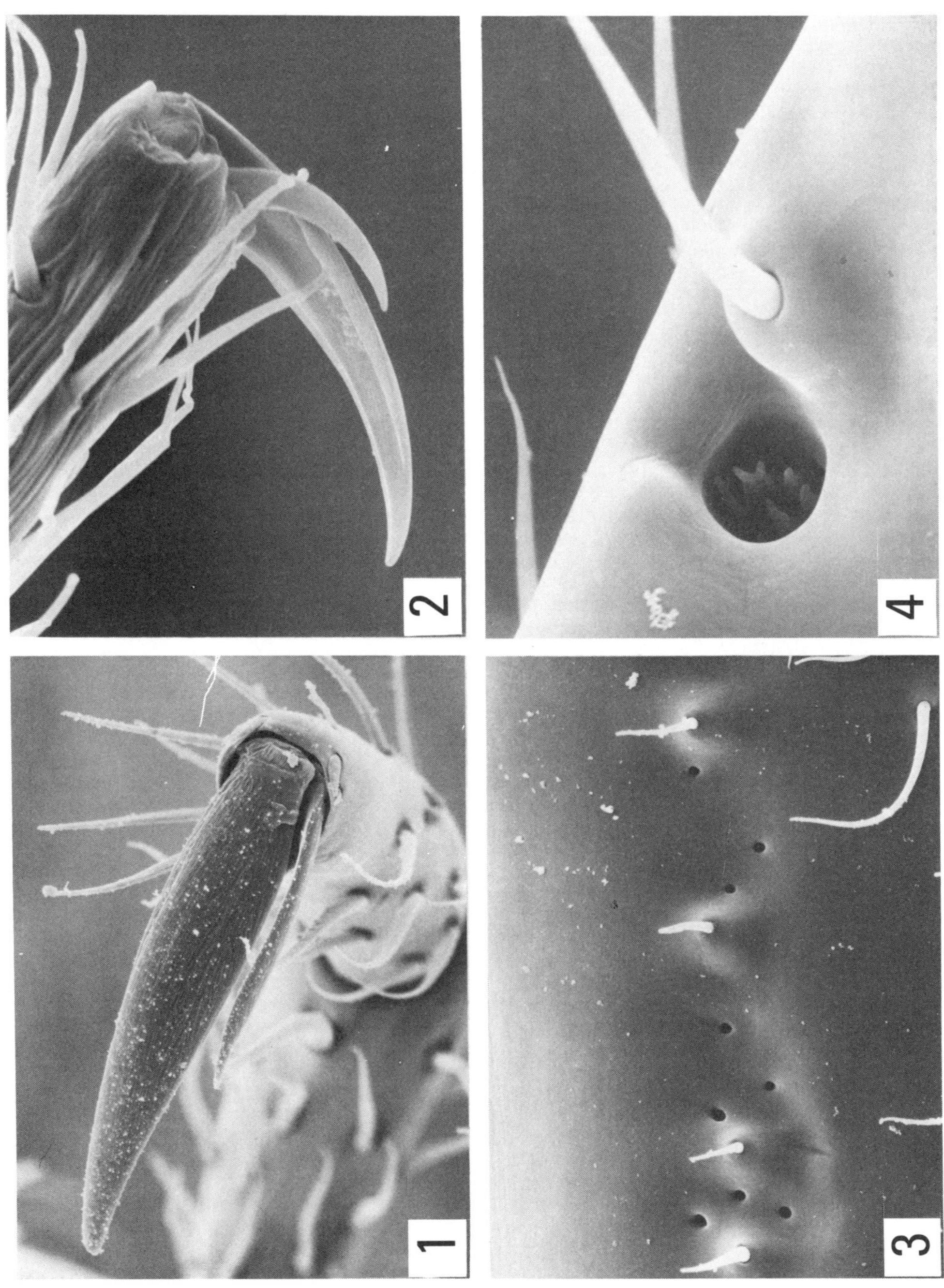

PLATE 79.

Fig. 1. Discal fovea, elytra, *Thesium cavifrons*, 5970X.
Fig. 2. Vertexal fovea, *Oropus basalis*, 4100X.
Fig. 3. Sutural fovea, elytra, *Melba* sp., 5750X.
Fig. 4. Sutural and discal foveae, elytra, *Oropus castaneus*, 1218X.

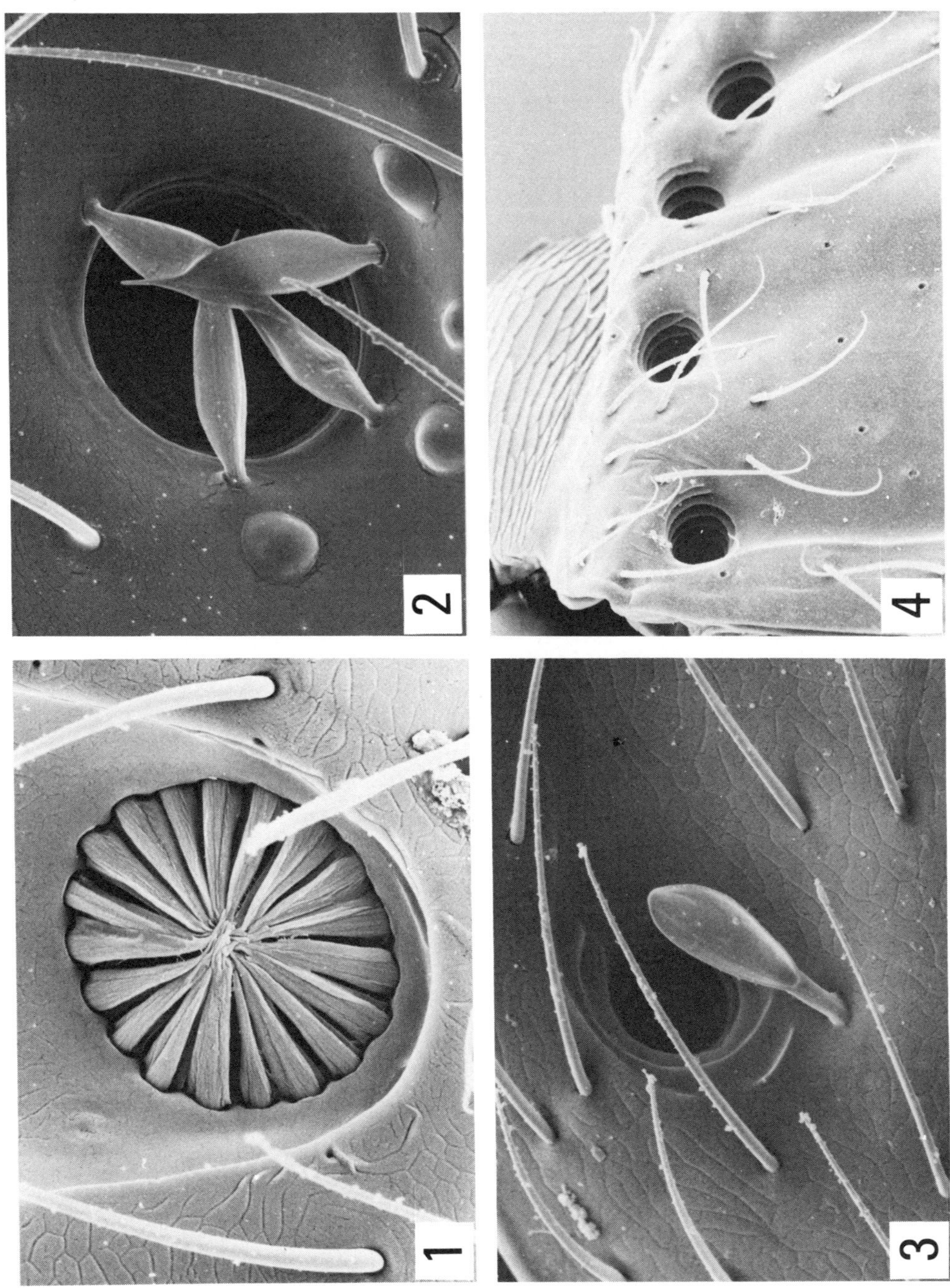